AUSTRALIAN ENVIRONMENTS

Noongar Country

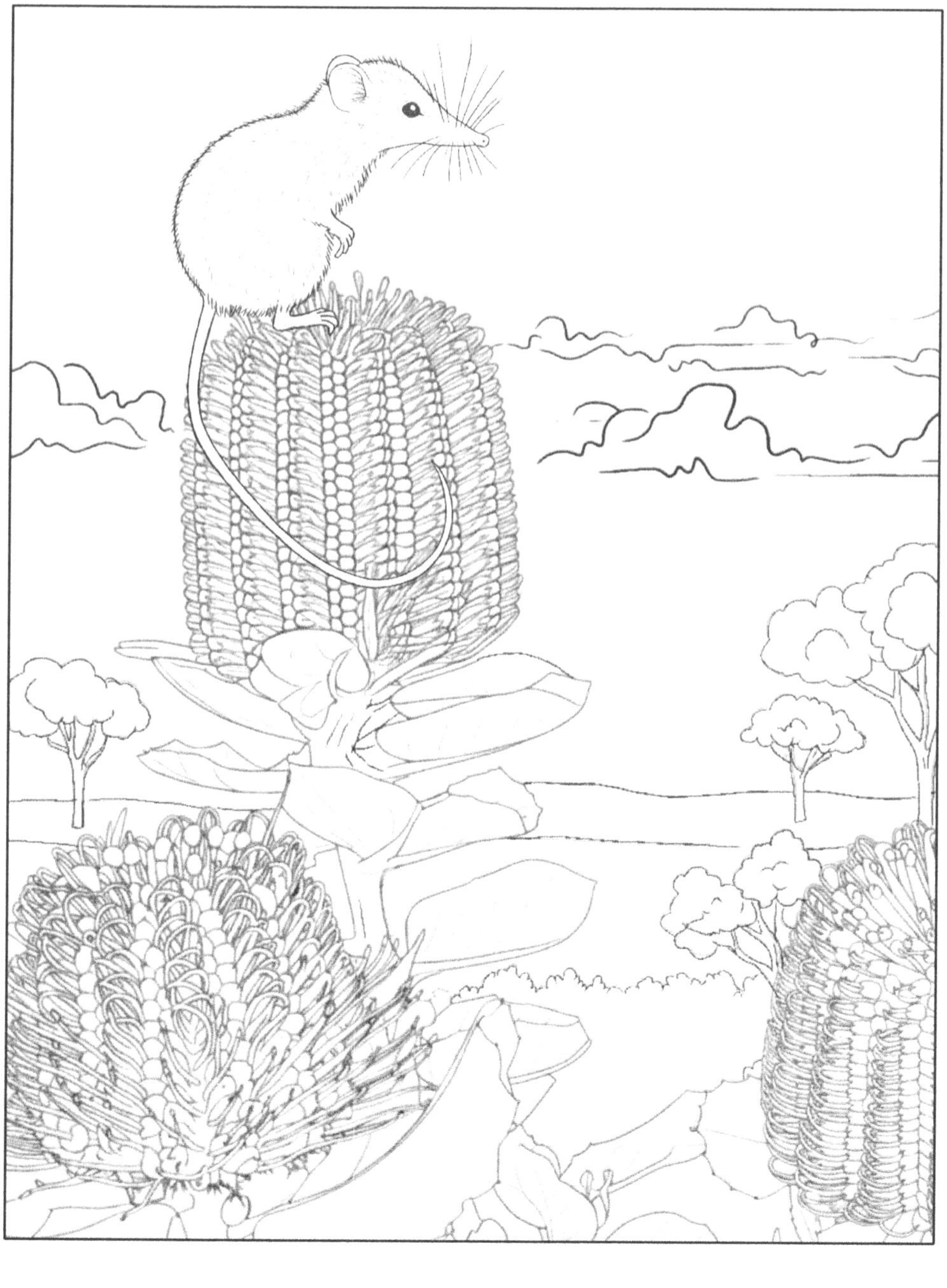

ISBN: 978-1-7641194-5-0 (Paperback)

First edition: November 2025
A CONSERVATION CLUB PUBLICATION
www.conservationclub.com.au

NATIONAL LIBRARY OF AUSTRALIA

A catalogue record for this book is available from the National Library of Australia

In the spirit of reconciliation, Conservation Club acknowledges the Traditional Custodians of Country throughout Australia and their connection to land, sea, and community. We pay our respects to their Elders, past and present, and extend that respect to all Aboriginal and Torres Strait Islander peoples.

To my grandchildren: Samuel, Jameson, Harry, and Audrey

.. in the hope that they will enjoy sharing this journey of discovery ..

"And, above all, watch with glittering eyes the whole world around you because the greatest secrets are always hidden in the most unlikely places. Those who don't believe in magic will never find it."

Roald Dahl

CONTENTS:

REWARDS:

Conservation Club offers rewards (such as free downloads of colouring sheets and button badges) which can be earned by answering the 20 Question Quiz on page 162 and submitting one coloured-in image with the colours being truly representative of the species concerned. We encourage keen eyes, writing, recording, and reporting species sightings.

- **Buy one of our pocket notebooks! Designed to fit into the pocket of a child at approximately A7 size, this blank notebook is an invaluable tool to have in order to record your observations of the environment you are exploring. It is really important to note the date, time of day, weather conditions and details of any sightings you may be lucky enough to experience! If the species is on the NCA, EPBC and/or IUCN Red List, please consider reporting your observations to an entity such as iNaturalist or Atlas of Living Australia! Citizen scientists are making great contributions to our understanding of these environments – you could be one of them!**
- **It is totally okay for you to guide your youngster and help them with this task (we suggest using the form at the back of the book as a guide before downloading the questions)!**

Email: conservationclubau@gmail.com for more information about how to apply for these rewards. The quiz questions are on the last page of this book. You can also join our mailing list to keep up to date with new releases.

Author's notes:

- Words and phrases shown within the book text in **BOLD** appear in the glossary/word lists at the back of this book.
- The whole point of the colouring pages is to encourage your young conservationists to engage with the book, tweak their observational skills, do a little bit of research, with the help of an adult if needed, and find out what colours are appropriate for the species concerned!
- Some of the words and language used may well be beyond what your young conservationist can understand, but it is suggested that you use the glossary to explain those words/phrases in simpler terms and at a level appropriate for your young wildlife warrior. They will ask for more information when they are ready!
- You can also choose to share the facts that you feel your young wildlife warriors are able to understand—you do not have to use them all! Just pick a few according to your child's age and comprehension skills.... You can always come back to the text later to expand their knowledge and vocabulary as they learn.

INTRODUCTION:

In these colouring pages, we focus on the **Noongar Country** environment and explore some of the species of animals and plants that call **Noongar Country** home. Some of the featured **species** are **critically endangered, endangered,** or **vulnerable;** the information for each species also includes this classification, where available. **Noongar Country** is located in the southwest of Western Australia, spanning an area from approximately Geraldton on the west coast to Esperance on the south coast.

There are 14 distinct groups of Australian Aboriginal people living within the Noongar cultural block and identifying as part of the **Noongar** community. These are: Amangu, Ballardong, Yued, Kaneang, Koreng, Mineng, Njakinjaki, Njunga, Pibelmen, Pindj arup, Wadandi, Whadjuk, Wiilman and Wudjari. The **Noongar** people refer to their land as ***Noongar boodja*** [2]

The **Noongar** peoples have six seasons, each defined by specific observable changes in the environment, with a dry period ranging from as few as three to as many as eleven months. Tribes are spread over three different **geological** systems: the coastal plains, the **plateau**, and the **plateau** margins; all areas are **characterised** by relatively **infertile** soil. The north is **characterised** by casuarina, acacia, and melaleuca **thickets**, and the south by **mulga scrubland**, but it also supports **dense** forest stands. Several rivers run to the coast, and lakes and **wetlands** provide the **Noongar** people with their distinctive food and **vegetation** resources.[2]

Generally, **Noongar** people made a living by hunting and trapping a variety of game, including kangaroos, possums, and wallabies. For those close to the coastal zone or **riverine** systems, spear-fishing or **culling** fish in traps was customary. An extensive range of **edible** wild plants were also available, including yams and wattle seeds. Nuts of the zamia palm, eaten during the *Djeran* season (April–May), require extensive treatment to remove their **toxicity**. While there has been some speculation that women may have used them as a form of contraception, there is no direct evidence to support this claim. As early as 10,000 before present, local people utilised quartz, replacing chert flint for spear and knife edges when the chert deposits were submerged by sea level rise during the **Flandrian transgression**.[2]

Before the arrival of Europeans, the **Noongar** population has been variously estimated at between 6,000 and some tens of thousands. **Colonisation** by the British brought both violence and new diseases, taking a heavy toll on the population. The **Noongar**, like many other Aboriginal peoples, saw the arrival of Europeans as the returning of deceased people, often imagining them as relatives who deserved accommodation. As they approached from the west, the newcomers were called *djaanga* (or *djanak*), meaning "white spirits".[2]

DID YOU KNOW?

Australia is the sixth largest country on Earth and the only country that covers a whole **continent**! Because of its **isolation** from other **continents**, many of Australia's animal and plant **species** are found nowhere else in the world. Australia also has a wide variety of environments, many of which are threatened for various reasons, including climate change, extreme weather events, and human impacts.

DID YOU KNOW THAT around 80% of Australia's native species are **endemic,** and many of them are thought likely to become **extinct** within our lifetimes? The focus of 'Conservation Club' is to try to raise awareness of **species** and environments that may be in our 'backyards', and yet, we may not be aware of them. Based on **fossil** records, the natural **extinction** rate is approximately one **species** per one million **species** yearly. Scientists estimate around 8.7 million **species** of plants and animals currently live on Earth. They are racing against time to catalogue Earth's **biodiversity** as **extinctions** continue. The Australian **continent** has seen more **mammal extinctions** than any other, along with significant losses of frogs, lizards, fishes, insects, and plants. Many of these **extinctions** are recent. In Australia, there have been 21 likely **extinctions** of animals and four of plants, averaging about four per **decade**. Out of the 23 **extinctions** for which causes have been identified, 17 (73%) were primarily due to **invasive** species. Focusing just on animals, 17 out of 19 (89%) **extinctions** were mainly caused by **invasive species**. The four plant **extinctions** are primarily attributed to **habitat destruction** or **degradation**. The primary **culprits** in recent extinctions include the ***chytrid*** **fungus** (affecting **amphibians**), which has wiped out six **species**, and the wolf snake, which is responsible for the loss of four **species**. Other significant impacts include cats, foxes, black rats, brown trout, and a ***trypanosome*** blood **parasite**. Although **habitat** loss is often **cited** as the primary cause of **extinctions**, this does not hold true for recent **extinctions** in Australia. Earlier conclusions linked many plant **extinctions** to **habitat** loss. Still, increased efforts in searching have led to the rediscovery of many of these plants. The current data indicates that **invasive species** have been the leading cause of most **extinctions** in Australia since **European settlement.**

LET'S EXPLORE!

Kids in the 8 to 14-year-old age groups are generally interested in spending time with their families exploring the great outdoors on weekend walks, camping getaways, etc. What a great way to spend time with your kids! So much better than burning up hours using electronic media and becoming **sedentary**, both of which can lead to behavioural problems and issues with sleep!

Kids also love to be part of a club and to start collecting things that they might be interested in, such as pebbles and shells. In my opinion, our kids are generally far more aware of **conservation** and **environmental** issues than in previous generations. This, therefore, presents an opportunity to educate, to get them interested, active, **motivated,** and involved in the **environment** and the **conservation** of **species** so that they will, hopefully, be able to work to preserve these animals enabling them to show or tell their own children about these special creatures and plants in years to come.

PLEASE.... GET OUT THERE and encourage these young and curious minds to engage with their local **environments** and the **critters** able to be found in them!

RESERVES, FORESTRY, NATIONAL & MARINE PARKS IN NOONGAR COUNTRY

- Ambergate Reserve
- Belvidere Beach
- Berridge Park
- Big Brook State Forest
- Big Pine, Bramley Forest
- Blackwood River National Park
- Booroora-Gardner National Park
- Boranup Karri Forest
- Bridgetown Memorial Park
- Brockman National Park
- Calgardup Cave and Leeuwin-Naturaliste National Park
- Canal Rocks
- Cape Naturaliste
- Central Lookout
- Crooked Brook Forest
- D'Entrecasteaux National Park
- Easter National Park
- Eyre Park
- Forest Path, Crooked Brook
- Georgiana Park
- Gloucester National Park
- Greater Beedelup National Park
- Greenbushes Heritage Park
- Gull Rock National Park
- Harewood Forest
- Hassell National Park
- Hawke National Park
- HMAS Swan Wreck
- Hoffman State Forest
- Jane National Park
- Kalgan River
- King George Sound
- Knoll Scenic Drive
- Kojonup Apex Park
- Kondil Park Recreation Area

- Lake Muir National Park
- Layman Picnic Site
- Leeuwin-Naturaliste National Park
- Lefthanders
- Leschenault Inlet
- Leschenault Peninsula
- Logue Brook
- Maidment Lakes
- Manea Park
- Minninup Piil
- Monastery Landing
- Mount Frankland National Park
- Mount Lindesay National Park
- Mount Roe National Park
- Myalup State Forest
- Ngari Capes Marine Park
- One Tree Bridge
- Porongurup National Park
- Powlalup Nature Reserve
- Preston National Park
- Rapids Conservation Park
- Scott National Park
- Shannon National Park
- St John Brook Conservation Park
- Stirling Range Entry Station
- Stirling Range National park
- Sugarloaf Rock
- Thornton Park
- Tone-Perup Nature Reserve
- Torbay Head
- Torndirrup National Park
- Tuart Forest National Park
- Two Peoples Bay nature Reserve
- Vasse River and Rotary Park
- Walpole and Nornalup Inlets Marine Park
- Walpole Nornalup National Park
- Wardandi Memorial Park
- Warren National Park
- Waychinicup National Park
- Wellington National Park
- West Cape Howe National Park
- Whicher National Park
- William Bay National Park
- Wiltshire-Butler National Park
- Wooditjup National Park
- Yalgorup National Park
- Yelverton National Park

NB: Whilst every effort has been made to include all parks, reserves, etc., this list may be incomplete, but there are certainly enough listed for many outings to explore the variety of environments **Noongar Country** has to offer!

Some background information about this book:

The species in this book occur in the ***Noongar Country environment****. The majority of the species featured are listed on the IUCN Red List and categorised as* ***Extinct, Critically Endangered, Endangered, Vulnerable, Near Threatened*** *or of* ***Least Concern****. Some of the species illustrated also appear on the EPBC (SPRAT) database and on the Atlas Of Living Australia website. A further source of information is iNaturalist. Species can be categorised differently across states and databases.* ***The IUCN Red List categories are those that have been used for this book****.*

On the IUCN Red List, more than 172,620 species have been assessed to date. However, of the goal of 260,000 species, at the time of publication of this book, 87,380 species remain to be evaluated —an incredible achievement —but, for some species, due to the lack of up-to-date scientific data, this list is, at this point, not complete.

The following tables provide information on how the species are listed under the EPBC Act (within Australia) and the IUCN Red List, including the categories they are listed in, and a breakdown of the categories currently available for assessment.

Younger readers may be unable to navigate this information on their own, so it is provided here to enable parents or caregivers to relay it to those young, curious minds!

I firmly believe that the age groups this book is aimed at (8 to 14 years) are curious, generally more informed than previous generations, and recognise that even a single voice, one person, can make a difference, as evidenced by the number of young people taking up a cause. For that reason, Conservation Club AU offers small rewards for engagement with this series. The current series will focus on Australian environments, but may well include other environments that are significantly important globally in the future.

To obtain a reward badge and for further information:
Email **conservationclubau@gmail.com** *or write to Conservation Club, PO Box 4909, Sunshine Coast MC, QLD 4560, AUSTRALIA.*

I sincerely hope that you, the Wildlife Warriors for the environments in Noongar Country, will enjoy this book as much as I have enjoyed writing and illustrating it!

Elysia Rainwood

EPBC – Status under the Environment Protection and Biodiversity Conservation Act 1999 [2]

EX	**Extinct**
XW	**Extinct in the wild**
CE	**Critically endangered**
E	**Endangered**
V	**Vulnerable**
CD	**Conservation dependent**

IUCN – Status on the IUCN Red List [1]

EX	**Extinct**
EW	**Extinct in the wild**
CR	**Critically endangered**
EN	**Endangered**
VU	**Vulnerable**
NT	**Near threatened**
LC	**Least concern**
DD	**Data deficient**
NE	**Not evaluated**

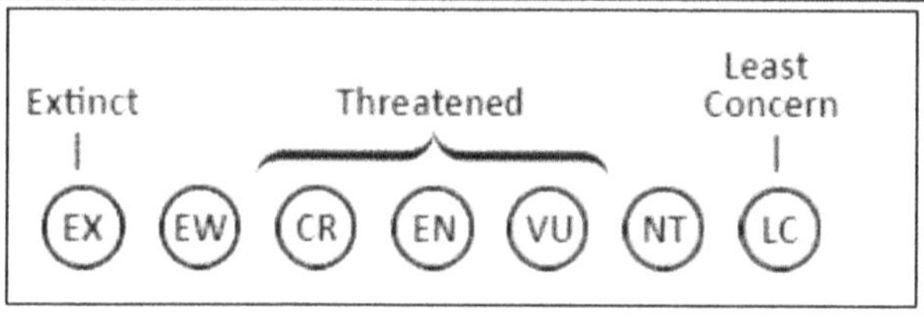

Where the species are included and categorised on the **IUCN Red List,** the category is outlined in bold and grey in colour!

IUCN categories explained: [1]

CE = Critically Endangered:

Facing an extremely high risk of **extinction** in the wild

EN = Endangered:

There is a very high risk of **extinction** due to rapid **population** declines of 50% to more than 70% over the previous 10 years, and current **population** size of fewer than 250 individuals.

V = Vulnerable:

Being threatened with **extinction** unless the circumstances that are threatening its survival and reproduction improve.

- Sometimes the species on the IUCN Red List have not been listed simply because there has not been sufficient research to assess.
- In these cases, they are regarded as DD = Data Deficient **OR** NE = Not Evaluated
- That does not mean that they are not **critically endangered, endangered, vulnerable, threatened** or any other classification... It just means that they have not been assessed!

Where on earth is Noongar Country found?

The shaded area indicates an approximation of the Noongar language region.
https://www.tandfonline.com/doi/full/10.1080/14443058.2020.1746380#d1e441

Noongar Country encompasses a variety of environments and ecosystems, including coastal plains, freshwater systems like rivers and lakes, forests (such as Jarrah and Karri), woodlands, and semi-arid regions. The landscape is rich with diverse flora and fauna, which sustained Noongar life and spiritual beliefs across different geographical zones and seasons.

NOTES:

- We recommend using colour pencils rather than texters or felt pens to ensure the information pages remain readable. Texters can bleed through the colouring side and, therefore, affect the readability of the fun facts.
- We encourage both adults and youngsters to find coloured images or photos online to guide the colouring in. Familiarising both yourself and them will assist in being able to recognise the species if sighted.

ABOUT THE AUTHOR

Meet Elysia Rainwood, an artist and writer with a passion for conservation and environmental protection, particularly the preservation of the diverse animals inhabiting our wonderful Australian continent.

Growing up amongst the rustling leaves of eucalyptus trees, Elysia developed a passion for the unique creatures and the diversity of plant life found in Australia. Elysia always had an eye for detail, being very aware of such things as seasonal changes in the various environments observed. From the depths of lush, green, sub-tropical rainforest to the islands of the Capricorn Bunker Group on the Great Barrier Reef, her childhood adventures fueled a desire to share the stories of the vulnerable and endangered.

Elysia writes non-fiction books and stories that resonate with the spirit of the tiniest of creatures and soar with the majesty of the wedge-tailed eagle. She likes to focus on the less well-known species, some of which may exist in your backyard, to ensure that they and their special habitats are not overlooked.

Elysia Rainwood's mission is clear:
To enchant young minds with the wonders of Australia's wildlife, fostering a generation of conservationists. Through the provision of interesting facts, engaging and reasonably accurate illustrations, she hopes to ignite a spark in the hearts of 8 to 14-year-old readers, inspiring them to protect the precious creatures that share our planet.

Along the way, Elysia hopes to also engage with parents, caregivers and educators who can pass on this information, as well as those who just enjoy the relaxation that colouring careful illustrations can bring.

So, whether you are a youngster or someone who shares this passion for conservation, join Elysia on this informative journey into the magical realms of nature, where words become wings for the endangered and vulnerable to soar.

Albany Banksia, Scarlet Banksia, Waratah Banksia, Waddib (Noongar)

Banksia coccinea

The Albany banksia is a unique Australian plant with some cool facts: it's a food source for animals and was used by Aboriginal people for **nectar**, it can survive bushfires, and its flowers are actually made of hundreds of tiny individual flowers. Many banksias, including the Albany banksia (a type of Coast Banksia), have a sweet, honey-like scent and are vital to the **ecosystem**.

- **Banksias aren't one flower:**
  A banksia isn't just one flower; it's a whole head made of hundreds of tiny flowers that grow together.
- **Fire is a helper:**
  For some banksias, such as the Albany banksia, the heat from a fire is necessary to open their woody fruit and release their seeds, making fire crucial for their growth.
- **Animals love them:**
  Birds, bats, and other animals love the sweet **nectar** inside the flowers, making them a super important food source in the bush.
- **Snack time for people:**
  Aboriginal people used to suck the sweet **nectar** directly from the flowers or mix it with water to make a tasty cordial.
- **They grow into trees:**
  Banksias can grow into tall trees, but they also grow as low shrubs. The tallest can reach up to 25m!

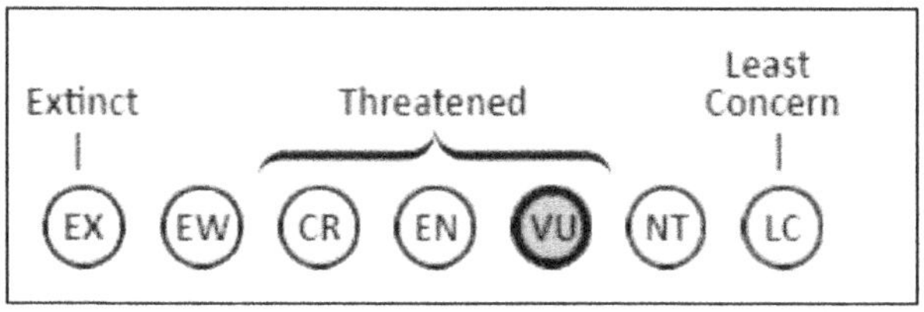

[1]

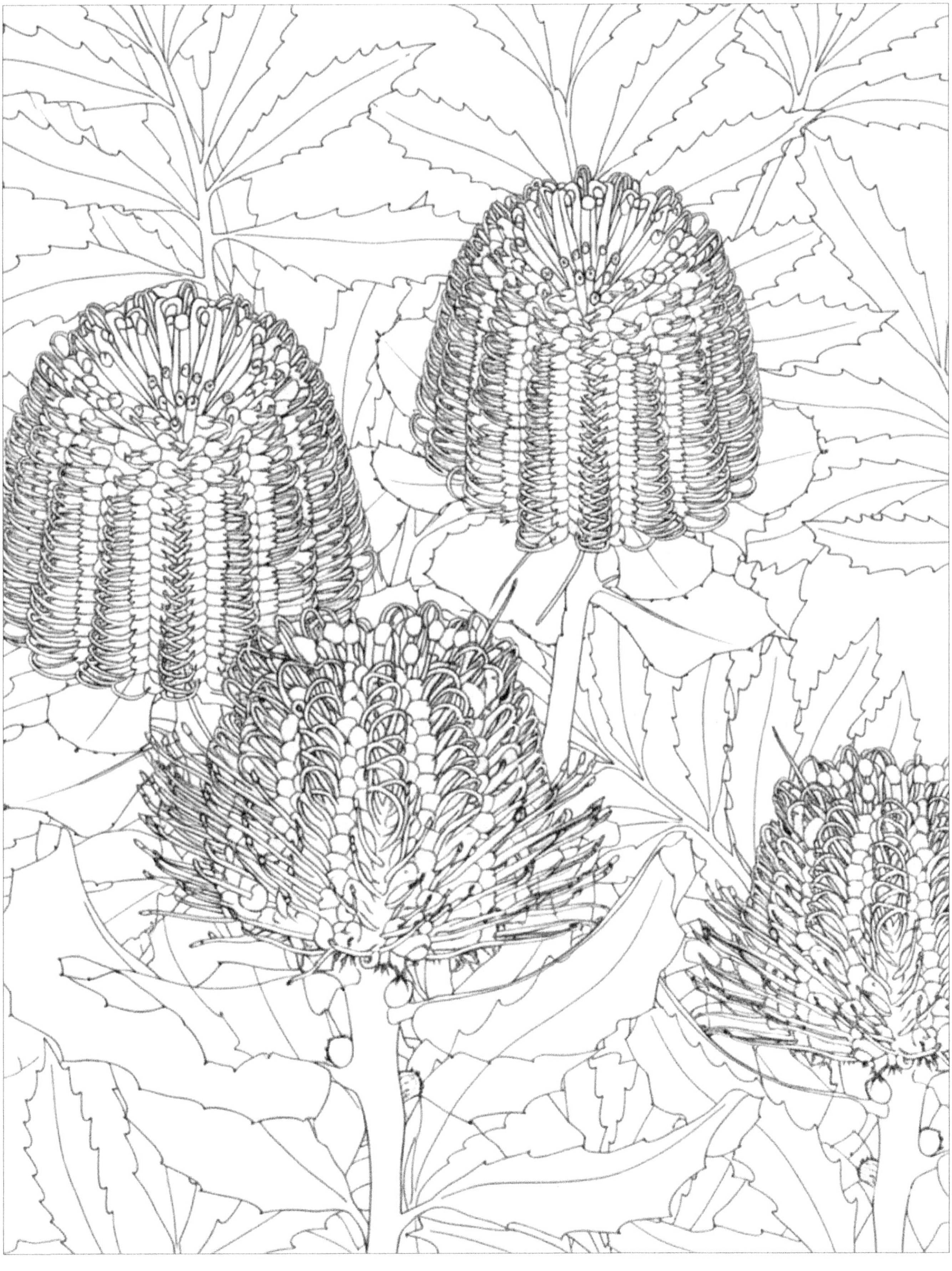

Banded Anteater, Numbat, Noombat (Noongar)

Myrmecobius fasciatus

Numbats are **unique marsupials** that are active in the daytime, which is unusual for their kind. They eat up to 20,000 termites a day using a long, sticky tongue and don't need to drink water. Numbats also have a long, bushy tail and distinctive white stripes on their back, and they sleep in hollow logs or burrows at night.

<u>Cool facts about numbats:</u>

- **Daytime sleepers:**
  Unlike most **marsupials** that are **nocturnal**, numbats are **diurnal**, meaning they are awake during the day.
- **Tiny eaters:**
  A numbat can eat up to 20,000 termites a day to get enough food and moisture.
- **Long, sticky tongue:**
  They use their excellent sense of smell to find termites and their long, sticky tongues to lick them up from tunnels and logs.
- **No pouch for babies:**
  Numbat babies are born tiny and cling to their mother's teats, which are located in a skin fold, for about six months.
- **Specialised home:**
  They sleep in hollow logs, tree hollows, or burrows at night to stay safe from **predators** and stay warm.
- **Striped fur:**
  Their reddish-brown fur has black-and-white stripes, which help them blend into their surroundings.
- **Endangered animal:**
  Sadly, the numbat is an **endangered** species because their homes are destroyed, and they are **preyed** upon by animals like foxes and cats.

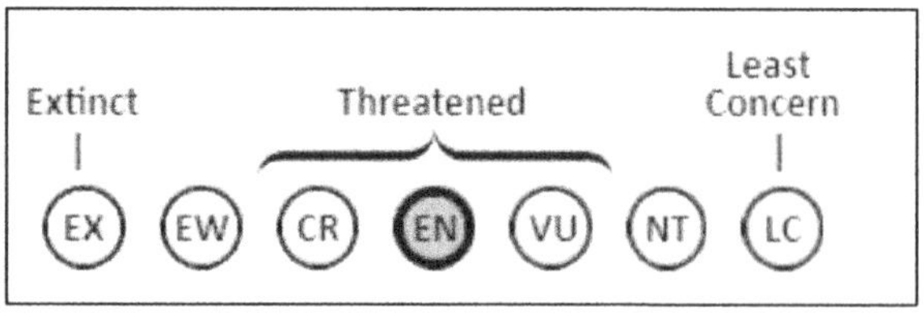

[1]

Sunset Frog

Spicospina flammocaerulea

The Sunset Frog is an ancient and colourful frog that lives in a small area of Western Australia. It's called the Sunset Frog because its bright orange and blue colours look like a sunset. Here are some interesting facts:

- **An ancient frog:**
  It is one of the oldest frogs in Australia, separated from its closest relatives more than 30 million years ago.
- **Unique look:**
  It has a dark purple-black back and a bright orange underside, with brilliant orange on its hands and feet. The back has many raised glands, giving it a spiny appearance.
- **Tiny size:**
  It's a small frog, with adults growing to be only about 3.5cm long.
- **Unique habitat:**
  It lives only in a few special peat swamps near the south coast of Western Australia.
- **A special song:**
  Male frogs make a unique call from late spring to the edges of pools and under dense cover.
- **Vulnerable:**
  Due to its small range and threats such as climate change, feral pigs, and wildfires, it is considered a vulnerable species.
- **A new discovery:**
  It was only discovered in the early 1990s and described as a new species in 1997.

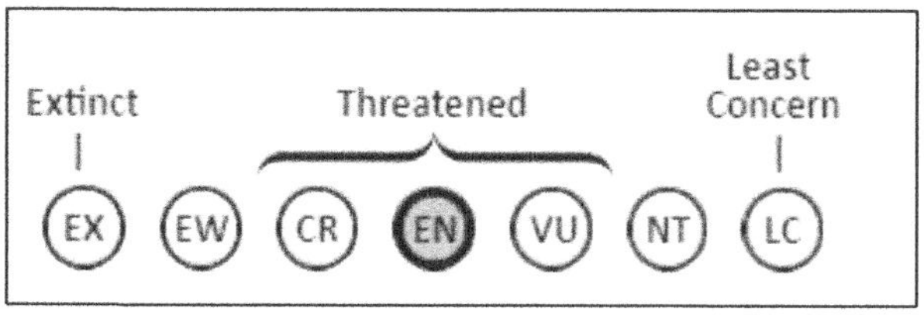

[1]

White-striped Octopus, Ornate Octopus, Djinda (Noongar)

Callistoctopus ornatus

The white-striped octopus is pink to red with white stripes and spots, and it lives on coral reefs. Like other octopuses, it is very smart, can change its colour to **camouflage** itself, and is a **nocturnal** hunter that eats fish and crustaceans. It can also use its eight arms to probe into **crevices** to find food and is capable of building a **den** with rocks for protection.

- **Appearance**
  It has a pink-to-red body with white spots down its arms and short white stripes on its **mantle** (the main body part).
- **Habitat and behaviour:**
  It lives on coral reefs and can be found **foraging** on reef flats or on sandy and gravelly bottoms, especially at night during low tide.
  It is an active night hunter.
  It builds its own **den** in rocks, sometimes creating a "door" to pull closed behind it.
- **Diet:**
  Its diet includes fish, **crustaceans**, and even other octopuses. It uses its arms to search for food in small holes and **crevices**.

Fun facts:

- **Super smart:**
  Octopuses are considered the most intelligent of all **invertebrates** and can solve puzzles and open jars.
- **Master of disguise:**
  It can change its colour to hide from **predators** or communicate with other octopuses.
- **Ink defence:**
  If threatened, it can shoot a dark ink cloud to confuse **predators** and escape.
- **Three hearts:**
  An octopus has three hearts: two pump blood to the **gills**, and one pumps it to the rest of the body.
- **Blue blood:**
  Their blood is blue because it is copper-based, not iron-based like ours.

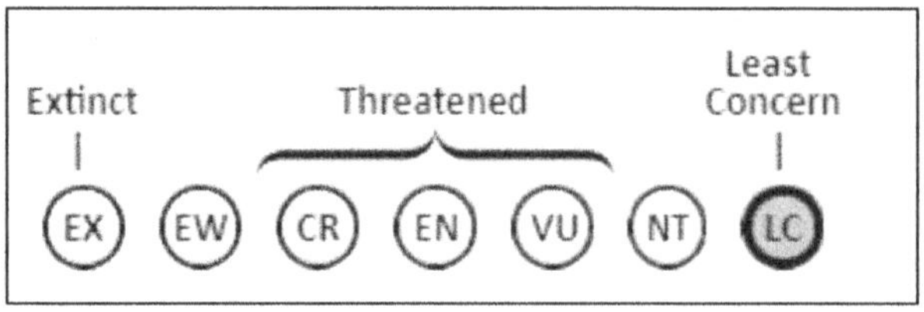

[1]

Black-tailed Godwit — Migratory

Limosa limosa

- There are four recognised **subspecies** of the black-tailed godwit, all with orange head, neck and chest in breeding **plumage** and dull grey-brown winter colouration, and a **distinctive** black and white wingbar at all times.
- Known for their long **migrations**, these elegant wading birds breed in wet grasslands and **bogs** in places like Iceland and Siberia, but they spend their winters in warmer areas like southern Europe, Africa, and Asia, **migrating** in **flocks**.
- Black-tailed godwits are tall birds with long legs that are great for wading in shallow water and a long, slender beak 7.5-12cm long, that they use to **probe** the mud for tasty insects, worms, and small creatures.
- In flight, its bold black and white wings and white **rump** can be seen readily. When on the ground, it can be difficult to separate from the similar bar-tailed godwit, but the black-tailed godwit's longer, straighter **bill** and longer legs are **diagnostic**. Black-tailed godwits are similar in body size and shape to bar-tailed godwits but stand taller.
- The most common call is a **strident** *weeka weeka weeka*.
- They are amazing travellers! They fly long distances from their breeding grounds to their wintering areas, often in large groups.
- These birds are very loyal. They often form strong bonds with their mates that can last for many years, returning to the same breeding grounds each spring to raise their chicks. Both the male and female black-tailed godwit work together to **incubate** the eggs and take care of the chicks after they hatch.
- A study of the Icelandic population showed that despite spending winter apart, pairs are reunited on their breeding grounds within an average of three days of each other. If one partner does not arrive on time, 'divorce' occurs. They nest in loose **colonies**. Unpaired males defend a temporary **territory** and perform display flights to attract a mate. Several nest scrapes are made away from the **courtship territory** and are defended from other godwits.[3]
- The species breeds in fens, lake edges, damp meadows, moorlands and **bogs** and uses **estuaries**, **swamps** and floods in the northern hemisphere winter; it is more likely to be found inland and on freshwater than the similar bar-tailed godwit.[3]

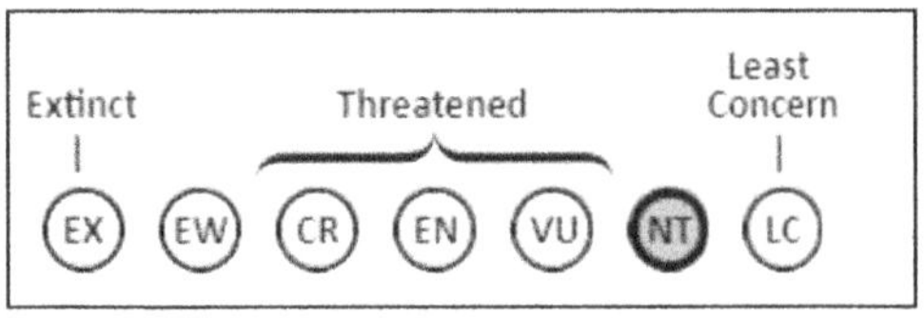

[1]

Tammar Wallaby, Tamar Wallaby, Kwoor/Kwaar (Noongar)

Notamacropus eugenii

Tammar wallabies are known for their unique ability to drink seawater when fresh water is scarce and for their **reproductive** strategy of nursing a joey in their pouch while a **fertilised embryo** develops in the **uterus**. They are **crepuscular**, spending their days resting in **dense vegetation** and feeding at dawn and dusk.

- **Unique reproduction:**
  A female tammar can have a **joey** in her pouch, another in her **uterus**, and a third suckling in the pouch. This allows her to be in a near-constant state of **pregnancy** and **lactation**, making her a key subject for research.
- **Seawater drinking:**
  The only **macropod** known to drink seawater, this **adaptation** allows them to survive in arid coastal areas like the Abrolhos Islands, although it comes at a **metabolic** cost.
- **Nocturnal behaviour:**
  They are most active at dawn and dusk, using tunnels in **vegetation** to move between their daytime shelter and nearby grassy areas to feed.
- **Vocal communication:**
  They communicate with each other through a variety of sounds, including hoarse coughs to defend territory and clucking during mating. They also use foot stamping to signal an alarm.
- **Historical significance:**
  The tammar wallaby was the first "kangaroo" to be encountered by Europeans when the shipwrecked crew of the Batavia spotted them in 1629 near the Wallabi Islands.
- **Physical traits:**
  They have thick, grey-brown fur, with males having reddish flanks. Their forearms are often held apart, particularly when moving.

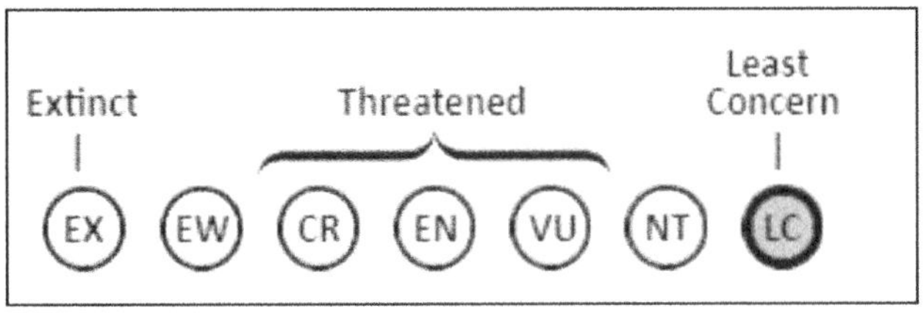

[1]

Leafy Seadragon, Glauert's Sea-Dragon, Mamoongat Ngoort (Noongar)

Phycodurus eques

The leafy seadragon is a master of disguise, using its leaf-like **appendages** to **camouflage** itself among seaweed and kelp to avoid **predators**. The male carries the eggs on his tail until they **hatch**, and these slow-moving fish lack teeth and a stomach, instead using a long snout to suck up small **crustaceans**.

- **Camouflage and appearance**
  Their name comes from the leaf-like **appendages** all over their bodies, which help them blend in with the surrounding seaweed and kelp. They can also change their colour to match their **environment**, a trait that allows them to **camouflage** themselves from **predators** like other fish. Their **camouflage** is so effective that even divers often struggle to spot them. Instead of scales, their bodies are protected by bony plates.
- **Diet and feeding**
  Leafy sea dragons do not have teeth or a stomach. They feed on small **crustaceans** and **plankton**, which they suck up through their long, tubular snouts. They have a specialised joint in their snout that snaps forward with incredible speed to create a **vacuum**, sucking prey into their mouths.
- **Reproduction and life cycle**
  Unlike most animals, it is the male leafy sea dragon that carries the eggs. The female lays eggs on a special "**brood patch**" on the underside of the male's tail, where they are **fertilised** and he carries them until they hatch. The eggs are attached to the male's tail and receive oxygen through specialised blood vessels in the **brood pouch**. Miniature sea dragons are born after about 6 to 8 weeks. As soon as they are born, the young are entirely on their own and must immediately start hunting for food.
- **Habitat and conservation**
  They live in **temperate** waters off the southern coasts of Australia, where they can be found on kelp-covered rocky reefs. The leafy sea dragon is the marine emblem of South Australia. They are threatened by habitat destruction, **pollution**, and **poaching** for the **aquarium** trade.

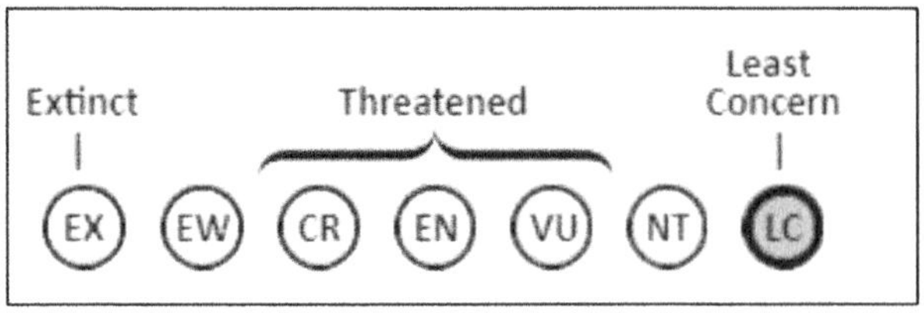

[1]

Australian pelican, Boodalang (Noongar)

Pelecanus conspicillatus

- Australian pelicans are one of the largest flying birds in the world (1.6 - 1.8 m long, 2.5 - 3m wingspan), only second to the snowy albatross (wingspan 3.5m), and can stay in the air for over 24 hours, travelling hundreds of miles as they soar in the breezes at heights of up to 3km and reaching speeds up to 56km per hour!
- Australian pelicans have an extremely light skeleton, which weighs only 10% of the whole weight of their body.
- Australian pelicans are known for having the longest bill of any bird in the world. Their massive throat pouch is used as a net to scoop fish and **crustaceans** and can also catch rainwater. It is so large that it can hold from 9 to 13L of water. There is a hook at the end of the bill that they use to hold slippery fish.
- They are bold black and white birds with blue legs, yellow eye-rings, and a pink or yellow pouch that turns bright red, blue, and orange during **courtship**.
- They are **opportunistic** feeders that primarily eat fish but also eat **crustaceans**, tadpoles, and even small turtles. They are known to steal fish from fishing nets or take scraps directly from fishermen, and can eat around 9kg of food a day.
- Pelicans are mouth-breathing birds. To cool off in hot weather, they pant like dogs to circulate air over their **saturated** pouch, which serves as a heat exchanger.
- A group of pelicans is called a "brief," "squadron," or "pod".
- They are found throughout most of Australia in both freshwater and saltwater **environments** like lakes, rivers, **estuaries**, and seashores.
- They live in **colonies** with thousands of other pelicans and often work together to catch fish and protect their young. The young leave the nest at around 28 days and form **creches** of about 100 chicks, which become independent at about 4 months, when their parents stop feeding them.
- The pelican family is at least 30 million years old!

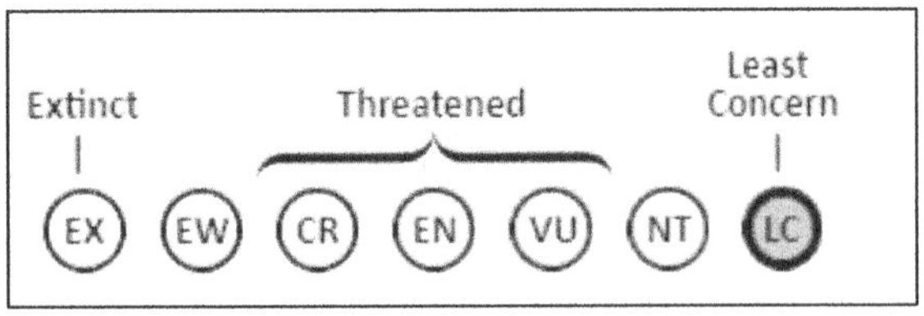

"A wonderful bird is the pelican...
His beak can hold more than his belly can.
He can hold in his beak
Enough food for a week!
But I'll be darned if I know how the hellican?"
Dixon Lanier Merritt

Pelicans DO NOT store food in the pouch of their bills!

*Free Printable Pelican Colouring Page – Simple Mom Project. https://simplemomproject.com/pelican-coloring-page/

Antipodean Fur Seal, Australasian Fur Seal, Black Fur Seal, Long-Nosed Fur Seal, Balgurt (Noongar)

Arctocephalus forsteri

Antipodean fur seals have excellent underwater vision, can dive for up to 15 minutes, and use specialised whiskers to hunt in the dark. They have a dual-layered coat for insulation, can recognise their pups by scent and call, and can move on land by using all four flippers.

- **Sensory and physical adaptations:**
  Antipodean fur seals have large, dark eyes that are adapted for seeing in the low light of the deep ocean. The long, sensitive whiskers on their snouts act like "feelers" to help them find prey in murky water. They have small external ears that are used for listening both above and below the water. They have two layers of fur—an outer layer of dark hair and a thick, light undercoat—to stay waterproof and warm. They can rotate their hind flippers forward, allowing them to "walk" on all fours on land, though they move more elegantly in the water.
- **Diving and movement**
  They can dive to depths of about 380m and hold their breath for up to 15 minutes, though they typically dive for much shorter durations. In the water, they can reach speeds of up to 9km/hr and can "porpoise" out of the water when moving quickly. While they appear awkward on land, they can move with surprising **agility**, using their front flippers to propel themselves forward.
- **Reproduction and social behaviour**
  A mother can find her pup among thousands by its unique call and scent. They also possess a unique **reproductive trait** called delayed implantation, which ensures pups are born in the summer, thereby increasing survival rates. Mothers go on **foraging** trips at sea, returning to the **colony** for a short period to suckle their pups. **Juvenile** males engage in **aggressive posturing** and fights, which can last for hours without a clear **victor**, as they practice **social** skills.
- **Diet and conservation**
  Their diet includes fish, squid, octopus, and **crustaceans**.
  They were once heavily hunted but are now a protected **species**, and their population is recovering.

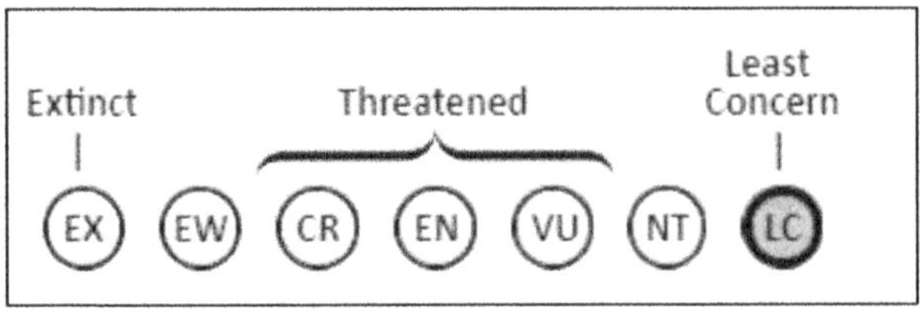

[1]

Gould's Goanna, Sand Monitor, Sand Goanna, Kaarda (Noongar)

Varanus gouldii

Gould's goannas, also known as sand goannas or racehorse goannas, are large, day-active lizards that are powerful diggers. They lay eggs in termite mounds, which provide a temperature and **humidity**-controlled **environment** for **incubation**. Recent studies suggest they are **venomous**, and their bites cause swelling and hinder blood clotting.

- **Diet and hunting**
  Gould's goannas are **voracious** eaters! They are strong **predators** with a varied diet of other **reptiles**, insects, small **mammals**, and **carrion**. They will also **scavenge** for food and are known to raid brush turkey nests for eggs.
- **Powerful diggers**
  Goannas have strong legs and claws used for digging their own burrows and for digging into other things to find food.
- **Habitat and behaviour**
  Goannas are primarily ground-dwelling reptiles that rarely climb trees. They dig extensive burrows for shelter and protection from the elements and **predators**, often hiding the entrance behind objects like logs or rocks.
- **"Stance" for viewing**
  When on the defensive or to get a better view of their surroundings, they will often raise themselves on their hind legs. They are known to wander through home gardens, especially in areas bordering large bushlands.
- **Breeding**
  Females lay their eggs inside termite mounds, and the termites then seal the eggs inside, providing a warm and humid environment. The eggs incubate for about 8 months before the young **hatch**. The **hatchlings** dig themselves out of the egg chamber and stay inside the mound for several days, feeding on termites before their first active emergence.
- **Venom and bite**
  Recent research confirms that monitor lizards, including Gould's Goannas, are **venomous**. Bites can cause immediate effects, such as swelling, localised disruption of blood clotting, and pain that can last for several hours. Bites are not typically deadly to humans, but they are painful and require caution. **Bacteria** in the mouth also increase the risk of infection.

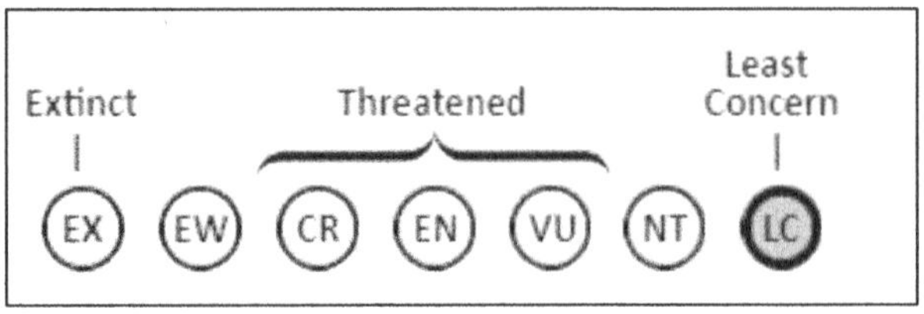

[1]

Banded Hare-wallaby, Mernine, Munning, Maning

Lagostrophus fasciatus

The banded hare wallaby is a vulnerable **species** native to Bernier and Dorre Islands off the coast of Western Australia. The species is also the only survivor of an ancient group of kangaroos and has been reintroduced to **predator**-free sanctuaries on mainland Australia at the Australian Wildlife Conservancy's Mt Gibson Reserve, as well as to Wadderin Sanctuary in the eastern Wheatbelt of Australia. Interesting facts include its "hare-like" **agility**, the ability for females to delay the development of a **fertilised** egg, and its **distinct** dark, **horizontal** stripes on its back.

- **Physical and behavioural traits**
  It is called a "hare" wallaby due to its speed and jumping ability, and it sits in well-formed, hare-like squats. It has a **distinguishing** series of dark, **horizontal** stripes that run from the middle of its back to the base of its tail. Its fur is long, thick, and predominantly grey, speckled with yellow and silver, and it has short ears and a short tail. It is active mainly at night, feeding on grasses and shrubs, and shelters in groups under bushes during the day.
- **Reproduction and development**
  Like other **macropods**, females can delay the development of a **fertilised** egg until the pouch is free, even if they have a **joey** inside. This means they can have a **joey** in the pouch and another **embryo** waiting in the pouch at the same time. This **reproductive** strategy is known as **embryonic diapause**. A female typically gives birth to one joey, which spends about 6 months in the pouch and is weaned at 9 months.
- **Conservation status and threats**
  Although the banded hare-wallaby was once found across the south-western portion of Australia, it is believed to have been **extinct** on the mainland since 1963, and the last recorded evidence of the banded hare-wallaby on the Australian mainland was in 1906.[4] The devastation of the **species** possibly can be attributed to the loss of habitat due to the clearing of **vegetation**, the loss of food (due to competition with other animals), and **predators**.
- Today, the only naturally occurring wild populations are found on Bernier and Dorre Islands, where they are protected.
- In 2013, the **species** was successfully reintroduced to Wadderin Sanctuary, and in 2017 to Mt Gibson Reserve, both **predator**-free **sanctuaries** on the mainland, creating new populations to protect it from **extinction**.

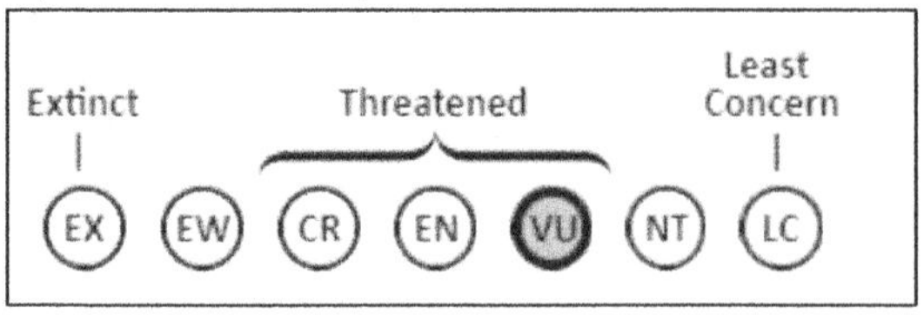

[1]

Green Turtle, Yakkin = turtle (Noongar)

Chelonia mydas

- Even though green turtles are very large as adults, they are only a tiny two inches when they are born!
- Green turtles spend most of their time underwater and can be there for up to five hours before coming up to breathe.
- Green sea turtles are among the world's largest turtles, weighing between 65 and 130kg and measuring between 1 and 1.2m long. Their strong, teardrop-shaped shell covers most of their body except for their head and four flippers.
- Why are they called "green" sea turtles? Because it is thought that their unusual coloured fat results from their vegetarian diet, green turtles eat marine plants such as seaweed and seagrass, unlike most other sea turtles. They were once extensively harvested for turtle soup but are now protected.
- These turtles nest along the coastlines of more than 80 countries worldwide, with the largest nesting populations in Costa Rica and Australia.
- Green turtles **migrate** long distances between their feeding grounds and mating/nesting sites, with recorded distances over 2,600 kilometres. Females usually travel thousands of miles from their feeding grounds back to the beaches they were born on every two to four years, whereas males may make the journey **annually**.
- As powerful swimmers, turtles propel through the water with strong paddle-like flippers. Although they prefer to cruise along at around 3km/h, if threatened, they can reach speeds of up to 35km/h.
- After mating in the shallow waters near the shore, the female comes ashore to a sandy beach, where she uses her flippers to dig a nest. She lays a **clutch** of approximately 115 eggs, covers them with sand, and then returns to the ocean. The temperature at which the eggs **incubate** influences the sex of the **hatchlings**. Below 27.7°C, the **hatchlings** will be male; above 31°C, they will be female; and between these temperatures, the **broods** will hatch both sexes.
- After about two months, the babies will use a special "egg tooth" to break their shells and hatch from their eggs. They dig themselves out and must cross the sand to reach the water, avoiding many hungry predators, such as birds, crabs, wild dogs, and lizards! A perilous journey.
- During the early years of a green sea turtle's life, they drift in the open ocean, feeding on **plankton**. As they mature, they **migrate** to coastal areas, including bays and lagoons, where they **graze** on seagrass.

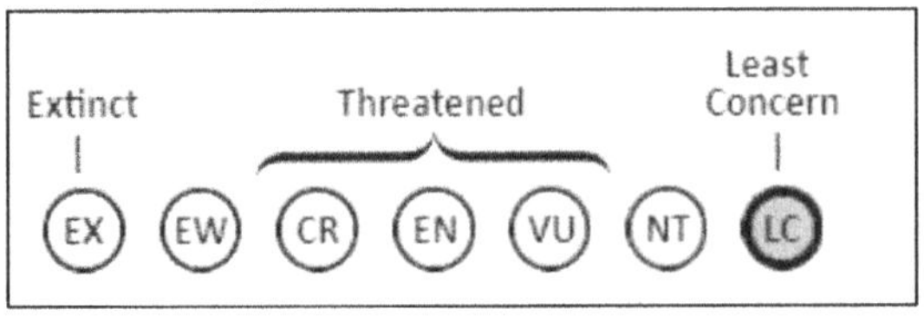

[1]

Carnaby's Black-Cockatoo, Short-Billed Black-Cockatoo, Ngoorlaark (Noongar)

Zanda latirostris

Carnaby's cockatoos are semi-**migratory** birds known for their **distinctive** appearance and their reliance on old-growth trees for nesting. They are social and live in large flocks, with males having a black beak and pink eye-ring, while females have a greyish-white beak and a paler eye-ring. They are nicknamed the "rain bird" because their breeding timing is linked to autumn rainfall, and their population is declining due to habitat loss and a low reproductive rate.

- **Social behaviour and migration**
  Carnaby's cockatoos are very social and can be found in pairs or large flocks, especially during the non-breeding season when they gather in large roosts. They migrate between coastal areas for feeding in the summer and the Wheatbelt for breeding in the winter, a journey of up to 12km one-way for feeding.
- Carnaby's cockatoos are known to form lifelong pairs.
- **Physical characteristics and identification**
  Males have a black beak and a pink eye-ring. Females have a greyish-white or bone-coloured beak and a greyish eye ring. Juvenile males look like adult females until they are about three years old. Both males and females have white cheek patches and a white tail band, but their beak and eye-ring colour are the best ways to tell them apart.
- **Breeding and lifecycle**
  Breeding occurs in the drier inland Wheatbelt during late winter and spring, with egg-laying timing closely following autumn rainfall. They nest in hollows of large, old trees, such as Salmon gum and Wandoo. They typically lay two eggs but usually raise only one chick due to food **scarcity**; however, both can be successfully raised if food is **abundant**. The male feeds the female during the **incubation** period and while the chick is very young; both parents then **forage** for food as the chick grows.
- **Other facts**
  They are sometimes called the "rain bird" because their breeding is closely tied to the timing of autumn rainfall. Carnaby's cockatoos have experienced a significant **decline** in their range and population due to **habitat** loss, mainly from widespread **vegetation** clearing. They are a threatened species, and their conservation is a major focus for organisations in Western Australia.

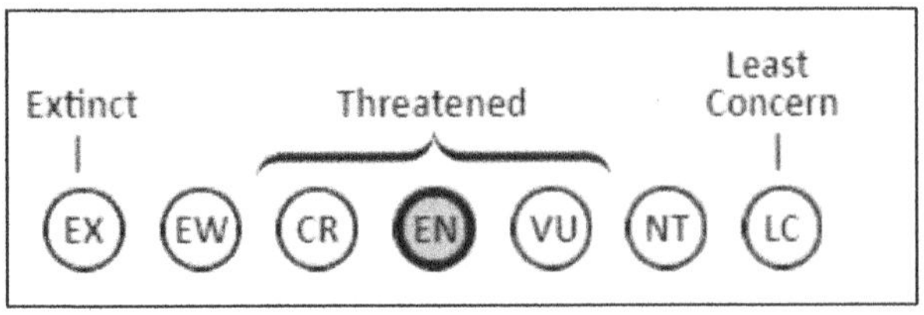

[1]

Spotted-thighed Frog

Litoria cyclorhyncha

The spotted-thighed frog is a large frog with a dark green to brown back and bright yellow spots on its groin and thighs. They are skilled climbers and jumpers, with unwebbed fingers and nearly fully webbed toes. They are known for being **voracious predators**, sometimes eating small mice, and have become an **invasive species** in parts of southern Australia where humans likely introduced them.

- **Appearance and physical traits**
  Females can grow up to 85 mm, while males are typically smaller at up to 65 mm. They have a dark green to brown back with bronze or gold patches and a very **distinctive** bright yellow spotting on the underside of their rear legs. They have unwebbed fingers and toes that are nearly fully webbed, with small discs.
- **Behaviour and habitat**
  They live in permanent **swamps**, ponds, dams, and sometimes abandoned **quarries** in southwest Western Australia. Breeding occurs in permanent water sources during spring and summer, with eggs laid in loose clumps. They are large, **predatory** frogs that eat a wide variety of prey, including mice and other frogs. Males call during the late spring and summer around the edges of pools.
- **Interesting facts**
  The spotted-thighed frog has been introduced to southern Australia, likely via human transport, and is now considered an **invasive species** there due to its **voracious** appetite and potential to outcompete native **species**. Scientists are concerned about its potential to spread further eastward and affect native **fauna** in places like the Murray-Darling Basin.
- **Mimicry**
  It is similar in appearance to the motorbike frog (Litoria moorei), though it is generally more vibrantly coloured and found further south and east.

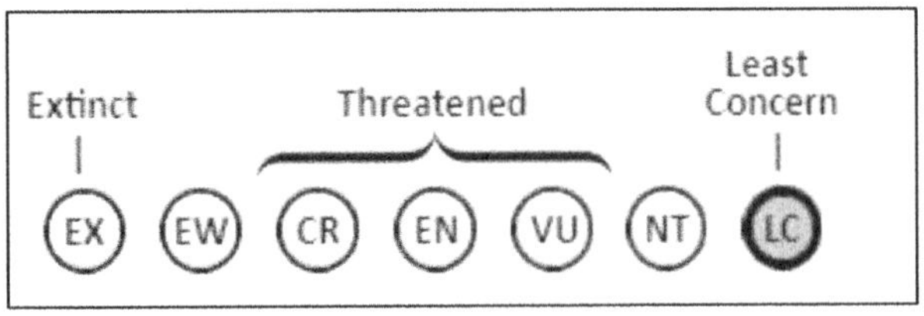

[1]

Red-necked Avocet, Yadjingoorong (Noongar)

Recurvirostra novaehollandiae

- The red-necked avocet, endemic to Australia, is an elegant shorebird known for its distinctive chestnut-brown head and neck, white body, and long, upturned black **bill**. It is a **nomadic** species that **inhabits** freshwater and saltwater wetlands and **estuaries** across mainland Australia, moving in small **flocks** in search of suitable **habitat**, especially in response to rainfall.
- Red-necked avocets are also known as the Australian avocet, cobbler, cobbler's awl and painted lady.
- Red-necked avocets feed by wading in shallow water and sweeping their upturned **bills** side-to-side just below the surface to find **aquatic** insects, **larvae, crustaceans**, and seeds. They will also sometimes swim and up-end to feed.
- These birds are highly **nomadic**, readily moving in small **flocks** to find suitable feeding grounds, especially after rainfall, and the availability of water in inland areas.
- During **courtship**, red-necked avocets engage in bowing dances and **bill**-crossing in shallow water, with males typically having a more sharply upturned **bill** than females.
- They breed in loose **colonies**, typically from August to September, although this can vary. Their nests are shallow scrapes lined with water plants in which they lay a single **clutch** of three to four eggs.
- Their call is often described as a "toot-toot" sound, similar to a clown horn or toy trumpet, and they also have sharper yelping notes for warning calls.
- Red-necked Avocets are generally **social** birds, found in loose **colonies** during breeding and small **flocks** outside of the breeding season, and they are known to associate with black-winged stilts.
- While relatively large, their eggs and chicks can be **vulnerable** to introduced **predators** like red foxes and feral cats.

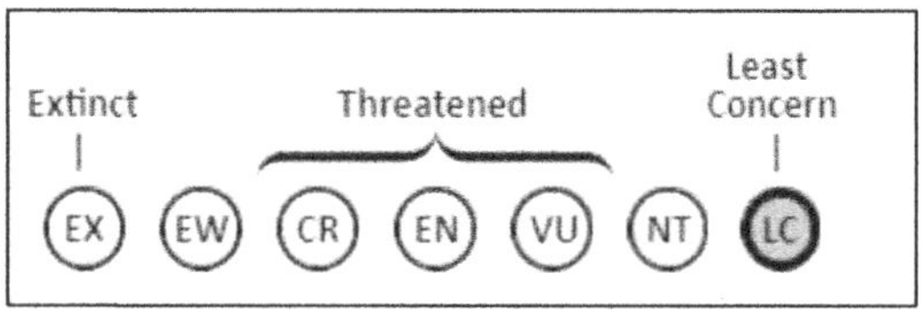

[1]

Sarah's Peacock Spider, Kara = spider (Noongar)

Maratus sarahae

Sarah's Peacock Spider is an interesting example of a tiny, colourful Australian jumping spider known for its **elaborate courtship** dances. Males display their brightly coloured **abdomens** by waving their third legs, and the females either accept or reject the dance, sometimes leading to the male being eaten. Unlike most spiders, they don't build webs and hunt their insect prey using excellent vision during the day.

- **Courtship and mating**
  The male peacock spider performs an **elaborate** dance to impress a female. This includes raising and expanding his brightly coloured **abdomen** and waving his third legs, much like a peacock fanning its tail feathers. The dance combines visual displays with **vibratory** signals. The female may eat the male if she is not impressed by his performance, if she is uninterested, or even after mating.
- **Appearance and size**
  Males are typically 2.5 to 6mm long, while females are slightly larger. Males are known for their bright, often **iridescent** colours, which are displayed through flaps on their **abdomen** during courtship. Females are usually a duller brown or grey colour.
- **Hunting and diet**
  They are active, **diurnal** hunters that **stalk** and jump on their **prey** rather than building webs. They feed on small insects, including flies, moths, ants, and other spiders. They have excellent vision, allowing them to spot prey and even see **ultraviolet light**.
- **Habitat**
  They are native to Australia and can be found in various habitats, such as bushland and domestic gardens. They can be found in low shrubs and leaf litter.

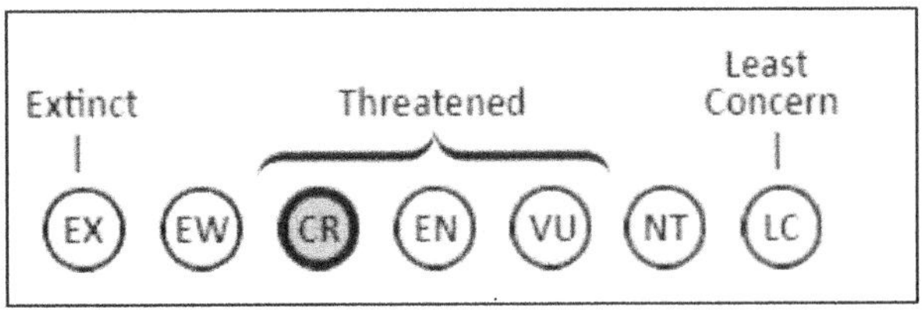

[1]

Shell-leaved hakea, Kwongan (Noongar)

Hakea conchifolia

The shell-leaved hakea is interesting for its unique, **rigid**, shell-shaped leaves that curl around the flowers. It is a small, **compact** shrub native to Western Australia, known for its blue-green **foliage**, and it is classified as "vulnerable" globally. Other interesting facts include its **lignotuberous** nature and its flowers, which appear from June to August and are white to pale pink.

- **Unique leaf structure**
  Its leaves are **rigid** and shaped like cupped shells, giving it the common name. The leaves fold along their **axis** to encircle the flowers. The leaves are blue-green, prickly, and grey-green when opened, with sharply scalloped edges.
- **Habitat and classification**
  It is found in heathland habitats between Perth and Dongara, extending inland to Eneabba. The IUCN Red List has classified *Hakea conchifolia* as a vulnerable species globally. It is a compact, **lignotuberous** shrub, meaning it has a woody root crown that stores food and water, allowing it to resprout after disturbances like fire. It is in the same group as *Hakea smilacifolia*, but is distinguished by being **lignotuberous** and producing more flowers per **inflorescence** (usually 15 - 18) compared to *Hakea smilacifolia* which has 5 – 8 flowers per **inflorescence**.
- **Flowering**
  It produces white to pale pink flowers in the winter, from June to August. Each flower cluster contains 15–18 flowers.

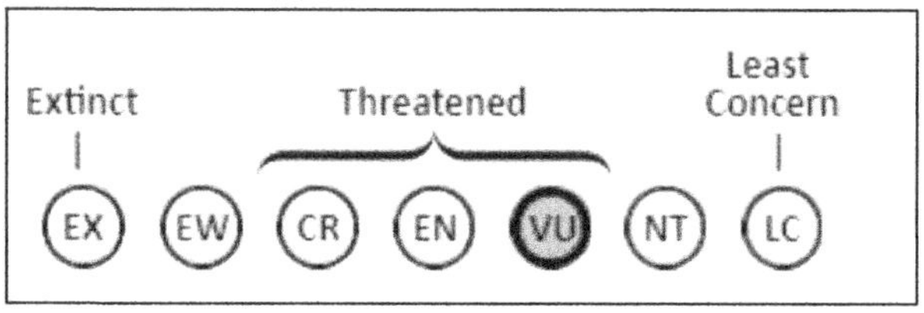

[1]

Thorny Devil, Mountain Devil, Muluwar (Noongar)

Moloch horridus

Thorny devils are known for their **unique adaptations**, including a "fake" head on their neck for **defence**, specialised skin with grooves that collect and funnel water to their mouths, and a diet of only ants. They also change colour to blend in with their surroundings, puff themselves up to appear larger, and have a unique, jerky walking style that helps them avoid predators.

- **Defence and appearance**
  They have a "**pseudo** head" on the back of their neck. When threatened, they tuck their real head between their front legs and present the false one to deter **predators**. The thorny spikes covering their bodies deter **predators** and help them collect moisture. Their colour can change to match their surroundings, providing camouflage. They can puff out their chest to appear larger and more **intimidating**. They have a slow, halting **gait** that involves rocking back and forth, which can help them blend in with their surroundings. The patterns on their bellies are unique to each individual, like fingerprints, and are used by scientists to study them.
- **Water and diet**
  Channels between their scales draw water droplets into their mouths via **capillary action**, allowing them to drink rainwater or dew that **condenses** on their body, and even absorb water from damp sand or puddles when rain occurs. They are specialist ant eaters and may eat up to 1,000 ants in a single meal. They use their short, sticky tongue to lick up ants they find on trails.
- **Behaviour and reproduction**
  They are active during the day. They are primarily solitary creatures. Males attract females with **elaborate courtship rituals** that include leg-waving and head-bobbing. Females lay 3 to 10 eggs in burrows between September and December.
- **Other facts**
  They can live for 15 to 20 years or longer.
  They tend to have separate, **specific** sites for **defaecation**.

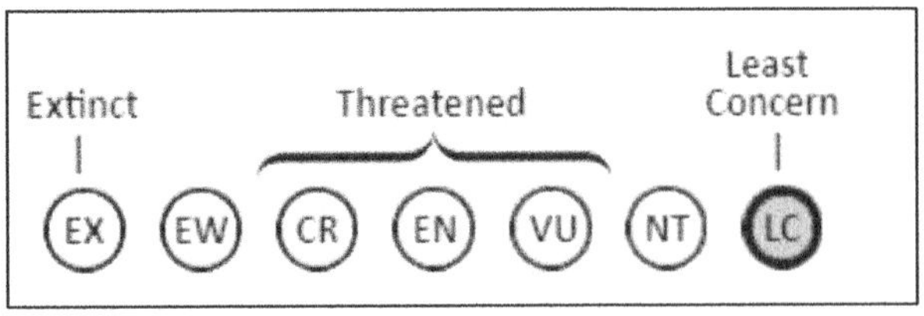

[1]

Watch a short video from @RemarkableAnimals on YouTube here:
https://youtube.com/shorts/KxIwjOTKdrA?si=BAAosRruX2TeltnC44

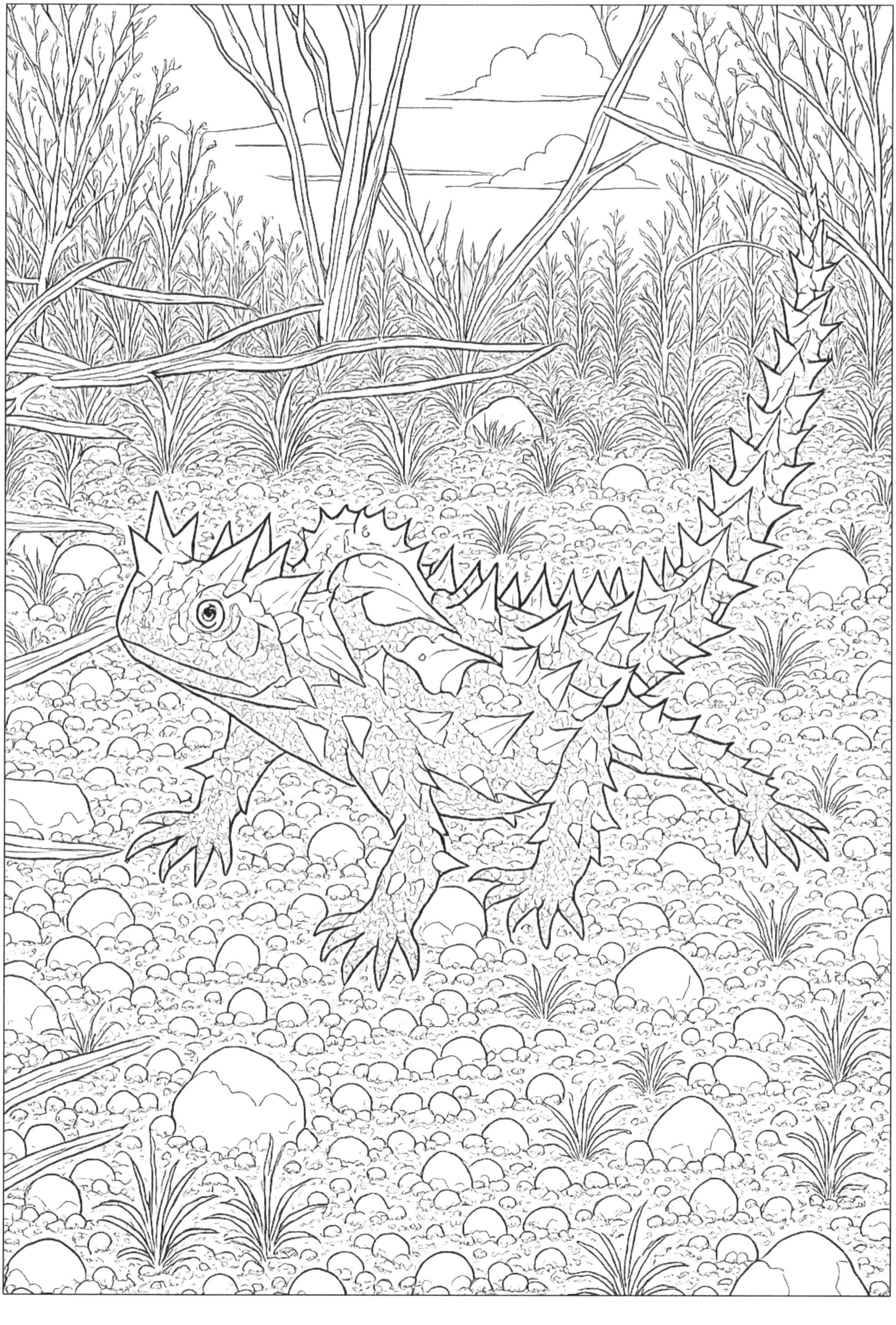

Mistletoe Bird, Moonidjedang (Noongar)

Dicaeum hirundinaceum

- The bottom-wiping mistletoe bird!
- The mistletoe bird is also known as the mistletoe flowerpecker, one of 54 species in the flowerpecker family, but is the only Australian example of the flowerpecker family.
- The mistletoe bird is highly adapted to its diet of mistletoe berries. It has a simple **digestive system** which the berries pass through quickly, **digesting** the fleshy outer parts, meaning that the mistletoe seeds are still sticky. The bird is **distinctive** in that it sits on twigs and 'wipes its bottom' on the twigs, spreading the seeds and leaving them with enough **organic material** to grow a new plant quickly. In this way, the mistletoe bird ensures a constant supply of its main food source, although may also catch insects mainly to provide food for its young.
- As the name suggests, these birds are mistletoe feeding specialists, although the mistletoe itself does not need the birds to spread its seeds as other fruit-eating birds, like the honeyeaters, also spread these.
- There are 92 species of mistletoe in Australia, more than anywhere else in the world, with about 70 being native and all 92 being endemic. Mistletoe is a **semi-parasitic** plant that needs its **host trees** for water but not necessarily for food. The plant can also **mimic** the appearance of the **host plants** making it very difficult to tell which plant is the mistletoe and which is the **host plant**!
- Mistletoe birds are thought to have started settling in Australia about 2 million years ago.
- The mistletoe bird is small; 9–10 cm long and 7.5–11 g in weight. Males and females have different markings and colours in their **plumage**. The male mistletoe bird could be mistaken for a scarlet robin (*Petroica boodang*) or flame robin (*Petroica phoenicea*). The mistletoe bird is **distinguished** by its lack of a forehead spot and a black streak on its white belly. Its **unique** combination of size, shape, behavior, and **plumage** makes it easily recognisable compared to other Australian **perching** birds, though it is closest in size and shape to the pardalote and thornbill. Their flight is fast, and they are usually seen flying above or high in the **tree canopy** on **distinctively** pointed wings.

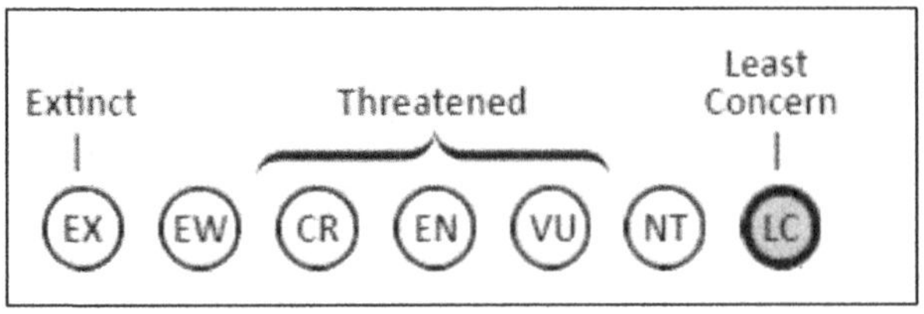

[1]

Western Ringtail Possum, Western Ringtail, Ngwayir (Noongar)

Pseudocheirus occidentalis

The western ringtail possum is an **arboreal marsupial endemic** to the southwest of Western Australia. It is distinct from other possum species and has several **unique characteristics**.

- **Physical traits and adaptations**
  Western ringtail possums have a **prehensile**, white-tipped tail. This long, strong tail helps the possum move through the forest canopy by grasping branches. It is also used to carry twigs and leaves for building its nest. They spend most of their lives in trees. The grasping forefeet are specifically adapted for climbing, giving rise to its scientific name, *Pseudocheirus*, meaning "false hand". To **navigate** and **forage** in the dark, the possum has large eyes **adapted** for nighttime vision. It has dark grey-brown fur, pale underparts, and shorter, rounder ears than its eastern cousin, the common brushtail possum.
- **Lifestyle and behaviour**
  Western ringtail possums build nests called "**dreys**," which are dome-shaped structures made of woven branches and leaves. They can use their tail to carry building materials to a new drey. A single possum may have between 2 and 20 **dreys** within its home range, moving between them throughout the year. They also use tree hollows and, in urban areas, may sleep in sheltered spots like roof spaces. These are relatively **social** animals that may share their nests and home ranges.
  The exact **gestation** period is unknown. A female typically has one **joey**, which stays in the pouch for about three months and remains dependent on its mother for up to seven months.
- **Diet and habitat**
  This possum is a **herbivore** (**folivore**), with a diet consisting primarily of the young leaves, shoots, flowers, and fruit from native plants. Its coastal populations heavily rely on the leaves of the Western Australian peppermint tree (*Agonis flexuosa*), which provides food and shelter. As they **forage** in the tree canopy, western ringtails **disperse** seeds and **pollen**, playing a **crucial** role in maintaining plant **diversity**. While historically found in forests, **urban** development has forced some populations to **adapt** to city life. They have been observed eating garden plants like roses and fruit trees.
- **Conservation status**
  The western ringtail possum is listed as critically endangered under Australian federal law, with populations having declined by over 95% since European settlement. Major threats include **habitat** loss and **fragmentation** from land clearing, **predation** by introduced foxes and cats, and the impacts of climate change. To help possums **navigate** urban spaces safely, projects have installed rope bridges to help them cross roads, a danger that has resulted in high **fatalities**.

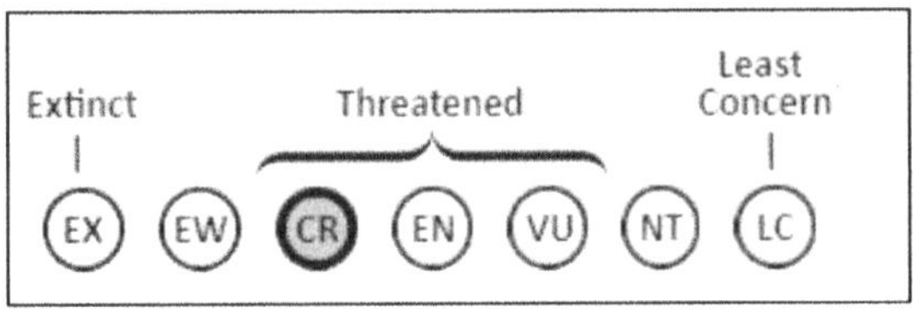

[1]

Ruddy turnstone

Arenaria interpres

- These birds are found along coastlines worldwide and **migrate** to breeding grounds in the Northern Hemisphere, then travel south for the non-breeding season, including to Australia from September to May. Ruddy turnstones get their name because they use their bills to flip over stones and seaweed to find yummy insects, **crustaceans**, and other small creatures hiding underneath.
- They are global travellers and **migrate** very far! Some Ruddy turnstones fly from Australia to Taiwan in just about six days, covering thousands of miles. They are capable of impressive non-stop journeys of up to approximately 7600km.
- They are amazing divers and strong fliers and can even dig holes in the sand to catch **prey** that's buried deeper.
- Their diet is varied, including insects, spiders, **crustaceans, molluscs**, and sometimes even eggs or **carrion**. They will eat maggots in rotting seaweed and sandhopper **larvae**.
- Ruddy turnstones have different looks for different times of the year. They are chunky birds with short orange legs and bills, and distinctive **plumage** that changes between breeding and non-breeding seasons. In breeding **plumage**, males have a bright reddish back, black chest patches, and a black and white face, while females and winter plumage are duller.
- Their calls include a **staccato** "tuk-a-tuk" or a warbling "tuk-tuk-tuk-tuk".
- Ruddy turnstones are **monogamous** and usually return to the same mates and breeding grounds each year. Both parents care for the chicks, though the female departs before **fledging**.
- Ruddy turnstone chicks are tough - they can leave the nest soon after **hatching** and are able to feed themselves, though they can fly after about a month.
- Outside of breeding season, they often gather in large groups, sometimes mixing with other types of shorebirds, but they can be quite **aggressive** when competing for food or defending their **territory** during breeding season.

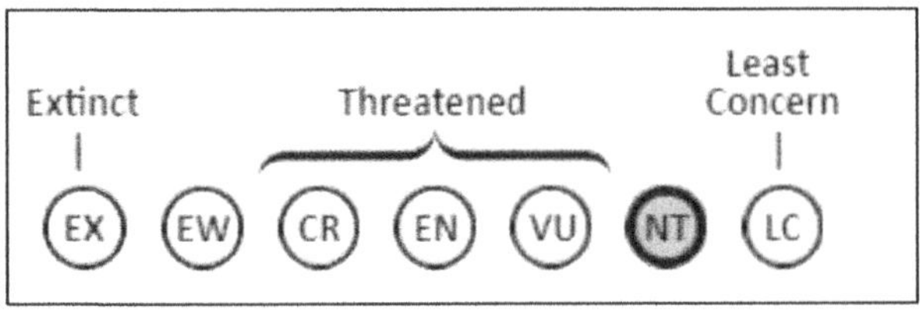

[1]

Australian painted-snipe

Rostratula australis

- An endangered bird found in the wetlands and **endemic** to Australia.
- Sightings of Australian painted-snipes are so rare that even keen birdwatchers with many years working in the field forget they exist, and no one has recorded the call of these **elusive** waders. Scientists are trying to learn where they go when they disappear, which they do for months and even years at times.
- When spotted, they seem to have a knack of arriving when there is plenty of water or just in time to enjoy rainfall in some regions. So, how do they know where to go? So many questions can be asked about this mystery species. They feed on wetland **invertebrates** such as worms, **molluscs**, insects, and **crustaceans**, as well as seeds and other **vegetation**.
- They have excellent **camouflage** and easily hide in the mud or amongst water plants around wetland areas. They can drop into dips and keep one of their highly positioned eyes on their surroundings.
- It is believed there are only about 340 individuals left, but there are other things that also make them rare; little is known about them at all!
- They have been described as the wetland equivalent of the night parrot, which is so **elusive** that for 100 years it was thought to be **extinct**!
- A study assessing the **evolutionary uniqueness** and **conservation** status of nearly 10,000 bird species worldwide ranked the Australian painted-snipe at number 29, highlighting them as rare and unusual birds.

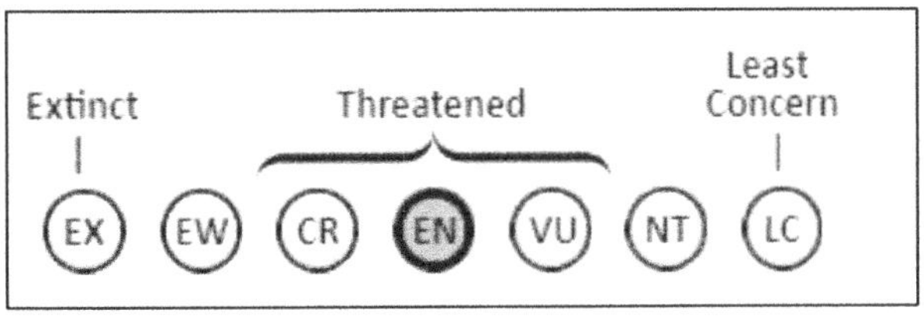

[1]

10,000 Birds:

Have a look at this list, it is so interesting, and has regular updates, as well as emails, focusing on different species of birds all around the world! Why not subscribe to their email list to learn more about all sorts of birds in all sorts of environments?!

- ❖ Here is the website address: (https://www.10000birds.com/about)

Short-beaked echidna, Nyingarn (Noongar)

Tachyglossus aculeatus

- Echidnas are **monotremes**, egg-laying **mammals**, in the same order as platypus.
- The echidna has spines similar to a porcupine, a beak resembling a bird's, a pouch like a kangaroo, and it lays eggs like a **reptile**! Also called spiny anteaters, these small, **solitary mammals** are native to Australia, Tasmania, and New Guinea. Typically, they measure between 30-43cm in length and weigh between 1.8-4kg.
- The echidna usually lives in open heathlands, forests, woodlands, scrublands, and grasslands, often among **vegetation** or inside hollow logs. During poor weather, they seek shelter under bushes or burrow into the soil.
- Echidnas have a long nose (sometimes called a beak), small eyes and tiny faces. They don't have very good vision, but have really good hearing and smell. They are also strong diggers, thanks to their long, rear-facing claws.
- Surprisingly, the spines on an echidna are an enlarged, tough form of hair; how about that?! As well as having spines, echidnas are covered in shorter fur that helps keep them warm. In colder climates, echidnas have fewer spines and thicker fur.
- Their spines also serve as the echidna's main **defence** against **predators**. If threatened, they roll into a ball with spines pointing outward or dig themselves to safety.
- Short-beaked echidnas eat ants and termites.
- Echidnas do not have any teeth. They have slender snouts and tiny mouths. They use their long and sticky tongues to feed on worms, termites, ants, and insect **larvae**. Their 15cm tongues rapidly dart in and out of their mouths to slurp up insects. They break down their food using hard pads located on the roof of their mouths and the backs of their tongues
- Echidna milk is pink because of the high iron content.
- Female echidnas do not have nipples but have special glands in their pouches that secrete the milk they feed to their young, and the baby echidnas called '**puggles**', lap it up.
- During the breeding season, echidnas form lines known as 'mating trains.' The male echidnas line up, nose to tail, behind a single female, forming a train of up to a dozen individuals.

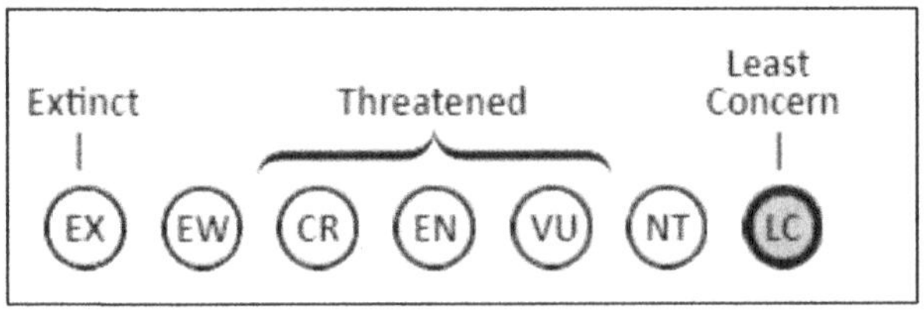

[1]

Shortheaded seahorse, Shortsnout seahorse

Hippocampus breviceps

The shortheaded seahorse is a master of **camouflage** known for its **elaborate** skin **tendrils** and vibrant, spotty colours. **Endemic** to southern Australia, this small species has surprisingly flirtatious mating behaviour that challenges the myth of seahorse **monogamy**.

- **Appearance and camouflage**
  The shortheaded seahorse has a tall, **prominent** coronet (crown) on its head and fleshy, frilly **tendrils** all over its head and back. Its colouration varies from drab grey to bright yellow-orange, purplish-brown, or reddish. Its body is covered in small, dark-edged pale blue or white spots called **ocelli**, which aid in **camouflage**. The shortheaded seahorse is a relatively small and slender species, reaching about 10cm in length.
- **Habitat and diet**
  This species lives in sheltered, shallow coastal waters up to 15m deep. It prefers seagrass meadows and areas with **abundant macroalgae** like Sargassum seaweed. It has also been found on floating **macroalgae**. During the summer, large numbers of **juveniles** can be seen clinging to floating **algae** on outgoing tides. The short-headed seahorse is a **carnivorous ambush predator**. With no teeth or stomach, it sucks tiny prey—primarily small **crustaceans** like **mysid shrimp** and **copepods**—through its tubular snout.
- **Social and reproductive behaviour**
  While many seahorse species are believed to be **monogamous**, researchers have found that short-headed seahorses "flirt" with more than one partner. They do not undertake the daily **social rituals** that reinforce pair-bonds in other species. Females have been observed to use a much larger range of territory than males. As with all seahorse species, the male gets **pregnant** and carries the developing young in a specialised **brood pouch**. After the female deposits her eggs, the male **fertilises** them. Males appear to give birth at night during the **Austral summer** months, typically releasing 50 to 100 young per brood.
- **Conservation status**
  The short-headed seahorse is **unique** to southern Australian waters, living in South Australia, Victoria, Tasmania, and parts of Western Australia. Its status highlights the importance of protecting its native **habitat**. More research is needed to determine population trends and connectivity. The species is protected in Victoria, and a ban on non-permitted fishing of **syngnathids** (the family that includes seahorses) is in place in Tasmania.

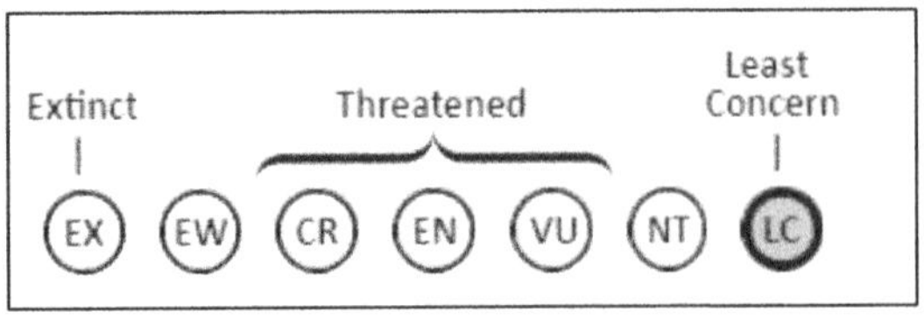

[1]

Burrowing Bettong, Boodie (Noongar)

Bettongia lesueur

The burrowing bettong, also known as the “Boodie”, is a unique marsupial known for its extensive underground **warrens** and role as an "**ecosystem** engineer". Once widespread across mainland Australia, it is now **extinct** there and survives naturally only on a few islands off the coast of Western Australia (Bernier Island, Dorre Island, Barrow Island, and Boodie Island).

- **Unique biology and social structure**
  The burrowing bettong is the only member of the kangaroo family (**macropod**) that creates and lives in large, **complex** underground **warrens**. The **warrens** can range from simple tunnels to vast networks with multiple entrances, which it excavates with its strong foreclaws. Bettongs are highly **social**, and their **communal warrens** can house between 20 and 40 individuals. Their social structure is believed to be **matriarchal**, with a **dominant** female, her daughters, and **subordinate** males occupying the **warren**. Burrowing bettongs exhibit **embryonic diapause**, a survival **mechanism** where a **fertilised embryo**'s development is paused. A female can have three young in her care at once: one recently **weaned**, one in the pouch, and a third in suspended **embryonic** development. Like other bettongs, the burrowing bettong uses its **prehensile** tail to carry bundles of nesting material like grasses and twigs.
- **Ecological role and impact**
  By constantly digging and turning over the soil in search of food, bettongs act as "**ecosystem** engineers" or "constant gardeners". This activity improves soil quality, increases water **infiltration**, and helps **disperse fungal spores** and seeds. While **omnivores**, bettongs are particularly fond of underground **fungi** known as **truffles**, which can make up a large portion of their diet. Their **foraging** plays a **crucial** role in distributing these **fungal spores**, which are essential for **nutrient** cycling and plant growth. In addition to **fungi**, bettongs eat a variety of other foods. Their varied diet includes roots, **tubers**, bulbs, seeds, and insects. During very dry periods, they will even eat the bark and roots of plants to get moisture.
- **Conservation and decline**
  The burrowing bettong was once one of Australia's most widespread **mammals**, but its population collapsed after European settlement. **Predation** by introduced feral foxes and cats was a primary cause of its **extinction** on the mainland. Today, the species survives in fenced, predator-free reserves and on a few isolated offshore islands. Conservation groups like the Australian Wildlife Conservancy are working to reintroduce them to safe **havens** in their former range. Research shows that bettongs can be taught to become more wary of **predators**. Projects at Arid Recovery in South Australia have shown that bettongs exposed to low cat **densities** can develop stronger anti-**predator** responses, with increased **wariness** and quicker escape **reflexes**.

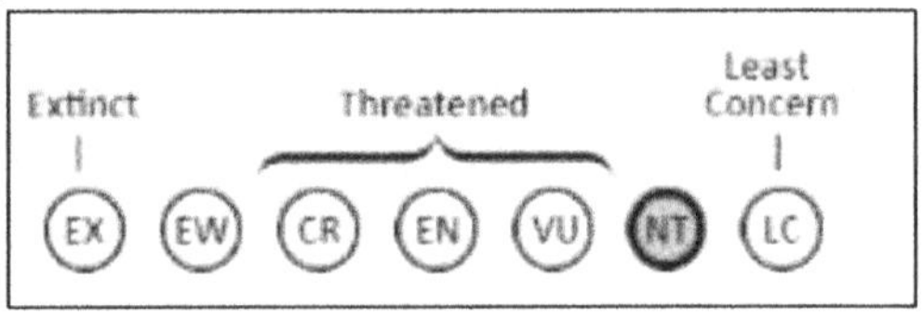

[1]

Western Bearded Dragon, Badart (Noongar)

Pogona minor (Subspecies: Pogona minor minor)

Western bearded dragons are known for their **unique** communication methods, such as puffing up their "beard" (which can turn black) and head-bobbing, to show **dominance** or **submission**. They are **omnivores** that can run surprisingly fast on two legs to escape **predators** or cool down, and they have a special third eye that helps them see what's behind them. Other interesting facts include their ability to change colour, their social nature, and that they undergo a winter state called **brumation** instead of true **hibernation**.

- **Communication and behaviour**
  They communicate through a variety of nonverbal cues, including head bobbing and arm waving. When threatened, they puff out and darken their spiky beards to appear larger, and head bobbing can indicate **dominance** or **submission**. They are **social** creatures that enjoy being around others and can learn behaviours by observing other dragons, a trait usually seen in **mammals** and birds.
- **Physical and sensory abilities**
  They can change their colour for **camouflage**, to regulate body temperature, or when they are feeling stressed or **aggressive**. They have a third eye, called the **parietal eye**, which is a small gap in their skull that allows them to see movement behind them. They can run surprisingly fast on two legs for short bursts, which helps them flee from **predators** and cool down by getting their bodies off the hot ground. Their strong jaws can crush hard-shelled insects like beetles.
- **Diet and lifestyle**
  They eat a varied diet of both plants and meat, including insects, flowers, fruits, and small **mammals**. They are comfortable on the ground and often **bask** on rocks and exposed branches in the morning and afternoon.
  They don't **hibernate** but enter a state called **brumation**, where they become less active for weeks or months, only waking up periodically to drink and eat.

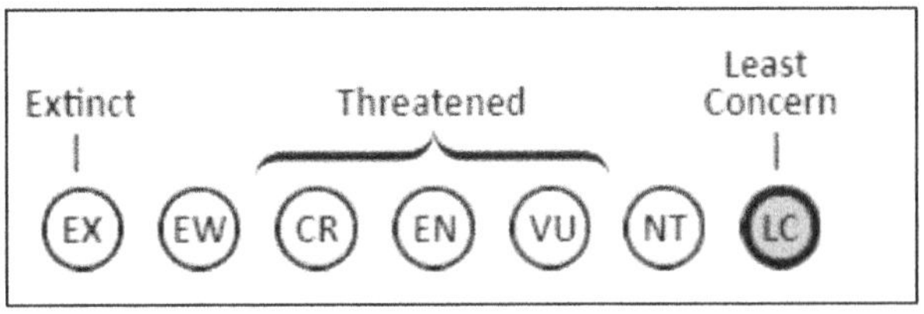

[1]

Freshwater crayfish species, Koonac

Cherax preissii

The koonac is a species of freshwater crayfish **endemic** to the southwest of Western Australia. While the term "koonac" is sometimes used generally for freshwater crayfish in the region, it specifically refers to the common koonac (*Cherax pressii*) and the glossy koonac (*Cherax glaber*).

- **Appearance**
  Koonacs can grow up to 20cm long and have broad, **serrated** claws. Their colouration can vary from black, blue-black, and reddish tones to dark brown, depending on their habitat. They are also identified by the four ridges, or "**keels**," on their head, with two being particularly **prominent**.
- **Habitat**
  These **crustaceans** are found in the wetlands, rivers, and seasonal waterways of southwest Western Australia. Unlike marron, which prefer more permanent waters, koonacs thrive in systems that dry up seasonally.
- **Burrowing ability**
  Koonacs are **adept** burrowers. This behaviour is a key survival **mechanism**, allowing them to dig into the **sediment** and stay connected to the water table for months at a time during dry summer months.
- **Diet**
  They are **omnivores** that feed on a variety of **organic** matter found on the streambed.
- **Conservation status**
  While the common koonac is rated as "least concern," local populations are threatened by competition for resources from the introduced eastern states yabby (*Cherax destructor*).
- **Related species**
  The glossy koonac (*Cherax glaber*) is a related species with a more restricted distribution in a small coastal region of Western Australia.

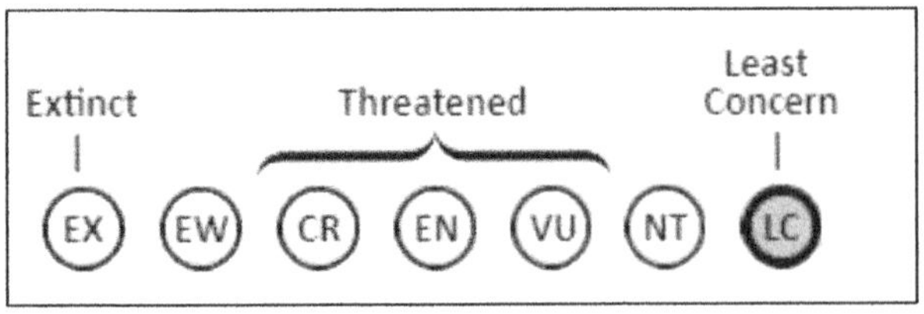

[1]

Eastern curlew

Numenius madagascariensis

The eastern curlew is the world's largest **migratory** shorebird, famous for its lengthy, **decurved bill** and an **epic annual migration** between its Siberian breeding grounds and its non-breeding **habitat** in Australia. The **species**, also known as the far eastern curlew, is considered critically endangered due to a severe population decline.

- **A remarkable migratory journey**
  Eastern curlews travel up to 12,000 kilometres one way, flying from breeding grounds in Russia and northeast China to coastal areas in Australia, with stopovers along the Yellow Sea. To fuel this enormous journey, the birds feast on crabs, **molluscs**, and insects on tidal mudflats, increasing their body weight by 40–70% before a continuous flight lasting two to three days. First-year chicks, just weeks old, have an **innate** ability to make their first **migration** without guidance.
- **Distinctive appearance and behaviour**
  As the largest wader in the world, the eastern curlew is easily identified by its extremely long, downward-curved **bill**. The **bill** of a female is typically longer than a male's. Its **plumage** is a flecked brown and tan, and it has a long neck and legs. It is a **wary** bird that is easily disturbed by human activity. This **wariness** is reflected in its large "flight initiation distance"—the point at which it will fly away. Its call is a **distinctive**, mournful whistle that sounds like "cuuuurlew," which is how the bird gets its name.
- **Critically endangered status**
  The global population of eastern curlews has decreased drastically, with an estimated decline of 80% over the last 30 years. The primary cause is the extensive loss of intertidal mudflat **habitats** due to coastal development and **land reclamation**, especially in the Yellow Sea region, a **critical** stopover point. Additional threats include human disturbance from **recreation**, **illegal** hunting, **pollution**, and the effects of climate change.
- **Conservation efforts**
  Because of its endangered status and **migratory** nature, conservation requires international cooperation along the entire East Asian–Australasian Flyway.
  A task force was established in 2015 to develop an action plan for the species.
  Conservation efforts in Australia focus on protecting **critical foraging** sites, raising public awareness, and working with international partners.

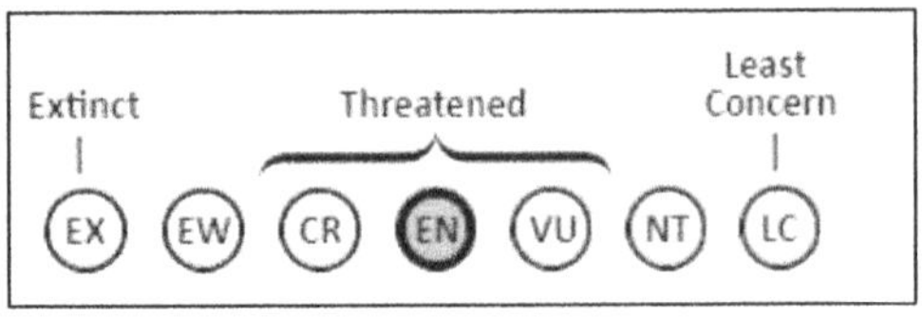

[1]

Crescent nailtail Wallaby, Waurong (Noongar)

Onychogalea lunata

The crescent nail-tail wallaby was an Australian **marsupial** declared **extinct** in the 20th century. Known for its silky fur and **distinctive** features, this small, shy creature had a hard, claw-like **spur** on its tail, which gave the "nail-tail" wallabies their name.

Key facts

- **Unique tail spur**: The wallaby's most unusual feature was a horny **spur**, or "nail," at the end of its tail, a trait it shared with its two close relatives.
- **Distinctive markings**: It was named for the light, moon-shaped (or crescent) markings that ran along its ashen-grey fur. It also had attractive stripes on its face.
- **Hiding in trees**: When pursued by **predators**, the **timid** wallaby would flee into a hollow log or tree. It would climb up the inside and sometimes emerge from an opening high above.
- **Awkward runner**: It was noted for its unusual running style, holding its short forelimbs awkwardly toward its chest.
- **Kangaroo rabbit**: English settlers nicknamed it the "kangaroo rabbit" due to its soft fur and long ears, which resembled an **exotic** rabbit.
- **Resembled and tasted like rabbit**: When hunted by early settlers for food, the wallaby's white meat was described as tasting similar to rabbit or chicken.
- **Indigenous names**: Aboriginal peoples of the central deserts knew the wallaby as "tjawalpa" and "warlpartu." The Nyungar people of southwestern Australia called it the "Waurong" or "Wurrung".
- **Extinction cause**: The crescent nail-tail wallaby was once common but declined rapidly in the early 1900s. Its **extinction** is believed to have been caused by a combination of factors, including **predation** by introduced red foxes and **feral** cats, as well as **habitat** loss from **pastoralism**.
- **Final sightings**: The last **specimen** to be collected alive was caught in a dingo trap on the Nullarbor Plain around 1927 or 1928. While some persisted in more **arid** areas until the 1950s, the **species** is thought to have become **extinct** around 1956.

[1]

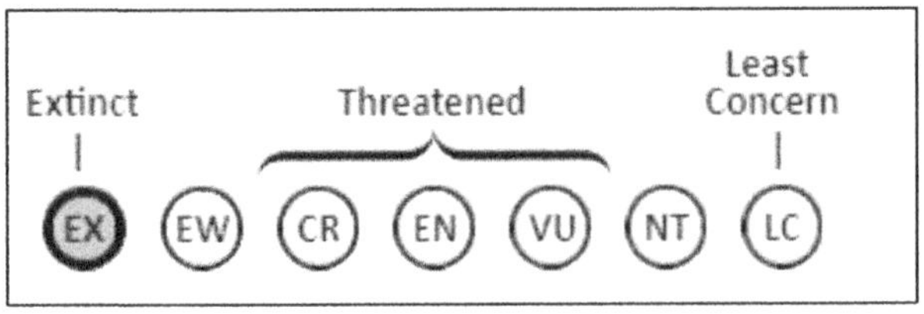

Rainbow Bee-eater, Birranga (Noongar), Birin-birin (Noongar)

Merops ornatus

- The rainbow bee-eater is a spectacular bird. With its green, blue, chestnut and yellow **plumage**, slim build, long and slender curved **bill** and **distinctive streamers** extending from the end of its tail, it is simply beautiful. It has a golden crown and a red eye set in a wide black stripe from the base of the **bill** to the ears, which is edged with a thin blue line. The throat is orange-yellow, with a broad black band separating it from a green breast.[5] The upper parts are green, with the flight feathers coppery and black-tipped. The underwings are bright orange with a black edge. The lower **abdomen** is blue. The tail, including the long **tail streamers**, is black with a blue tinge. Females have shorter, thicker **tail streamers** than males but are otherwise similar. Young birds are duller and greener, lacking the black band on the chest and the long **tail streamers**.
- There are many **species** of bee-eater in Africa, but the rainbow bee-eater is the only representative of the family in Australia.[5] It is a native bird in New South Wales protected under the Biodiversity Conservation Act 2016.
- Rainbow bee-eaters are often heard before they are seen. They are known to be noisy and **conspicuous**. The rainbow bee-eater makes a series of loud, **melodious** "pir-r-r" calls, **characterised** by rapid vibrating and high pitch, usually in flight. A softer, slower call is exchanged between **perching** birds.
- Rainbow bee-eaters eat insects, mainly bees and wasps, but also dragonflies, beetles, butterflies, and moths. They catch flying insects on the wing and carry them back to a **perch** to beat them against it before swallowing them.[5] Rainbow bee-eaters are **immune** to bee and wasp stings but will still rub them against the **perch** to remove the stings and **venom glands**. They can spot a potential meal from 50 metres away!
- Rainbow bee-eaters dig **burrows** in sandy banks/coffee rock to build their nests. The **burrows** can be over a metre long. While not **endangered**, the rainbow bee-eater's nesting period is **critical** to its survival. The **breeding** season for rainbow bee-eaters in NSW is from early November to January. Young birds leave the nest around February. Both parents **incubate** the eggs, sometimes with help from other bee-eaters in the **colony**.
- Bee-eaters have one of the most **complex social** systems of any bird. Many **species** (but not all) are **monogamous**, co-operative **breeders** living in large **colonies** of over 100 birds, which can reach 200 birds if conditions are favourable. Bee-eaters of both sexes alternate between being **breeders** and helpers in their **clan.**
- Threats to the rainbow bee-eater include **predation** by dingos, cane toads, monitor lizards, yellow-footed antechinus, and foxes. Dogs can dig up sandy nesting tunnels, and **human activity**, such as development and **quarrying**, disrupts their nesting.

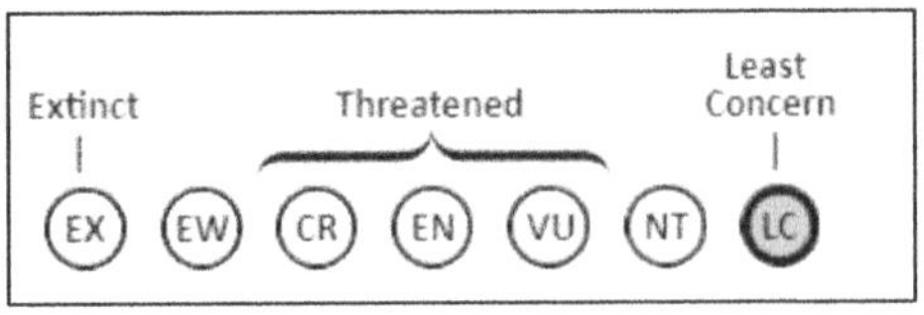

[1]

Greater Bilby, Dalgyte (Noongar)

Macrotis lagotis

The greater bilby is an Australian **marsupial** known for its long ears, digging abilities, and **unique adaptations** to **arid environments**. These **nocturnal** animals don't drink water, relying on the **moisture** from their diet of insects, seeds, fruits, and bulbs, and their backward-facing pouches keep them clean while digging extensive burrows.

- Bilbies are found in dry, desert areas and dry forests of Australia and are well-suited to hot, **arid** conditions.
- Bilbies have a pointed snout, long pinkish eats, a long black tail with a white tip, and strong forelimbs and claws for **burrowing** and digging up food.
- Males are much larger than the females and can be up to 2.5kg in weight.
- Each bilby moves between up to 12 burrows in their home range in response to the availability of food. The burrows may be 3m long and spiral down about 2m.
- They are strictly **nocturnal** so active at night, spending their days sheltering from **predators** and heat in their burrows and emerging at night to find food.
- Bilbies don't need to drink water because they get all the moisture they need from the food they eat.
- Bilbies usually live on their own, but sometimes form pairs or small groups including the young. Female bilbies have pouches that face backwards, preventing dirt from getting in when they dig.
- They have very good senses of smell and hearing, and retreat to their burrows if alarmed. Their large, hairless ears help them have excellent hearing to detect **predators** and also help them cool down in the heat.
- Bilbies have poor eyesight and rely more on their sense of smell and hearing to find food and sense danger.
- They feed on seeds (which are licked up by their long, slender tongues), fruit and insects (e.g. termites, ants and beetles), insect **larvae** (e.g. witchetty grubs), bulbs and **fungi**. Areas where bilbies are active may be pockmarked with shallow holes, up to 10cm deep, where they have been digging for food.
- In Australia, chocolate bilbies are often used in Easter baskets instead of Easter bunnies.
- The massive amount of soil bilbies turn over while digging helps the environment, making them important "**ecosystem** engineers".
- Unfortunately, bilbies are a vulnerable species in the wild due to factors like introduced **predators** like foxes and feral cats.

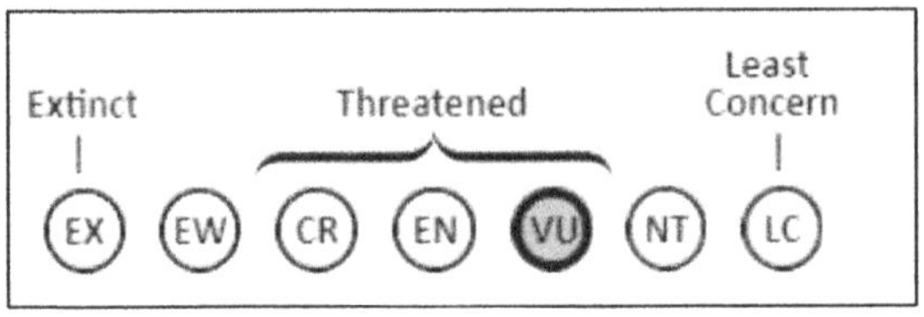

[1]

Western Quoll, Chuditch (Noongar)

Dasyurus geoffroii

Interesting facts about the western quoll include that it is a large, solitary **marsupial predator** with white spots on its brown fur and is a **nocturnal** hunter. They have experienced significant historical range reductions but are making a comeback, and they are known for defending their large home ranges. They are also excellent climbers, hunt a wide variety of prey, including small mammals, birds, and insects, and seek shelter in burrows or hollow logs during the day.

- **General facts**
  They have **rufous**-grey fur with white spots on their body and legs, but not their tail. They are about the size of a domestic cat and are the largest **endemic carnivore** in Western Australia.
- **Habitat and behaviour**
  They are **nocturnal** hunters, active at night, and hunt on the ground for a variety of **prey**. They are **solitary** animals that live in large home ranges, which they defend **vigorously**. They shelter in hollow logs, burrows of other species, or holes in **termitaria** during the day.
- **Diet**
  They are **opportunistic carnivores**, eating **mammals** (including rabbits), birds, lizards, frogs, insects, and carrion. They are excellent climbers and will also hunt prey in trees.
- **Reproduction**
  They breed seasonally, typically between late April and July. A female can give birth to two to six young. The young are carried in a shallow pouch for about 17–23 days, after which they are moved to a **den** to develop further. They become independent and can begin breeding in their first year.
- **Conservation**
  They were once widespread across mainland Australia but their range has contracted dramatically. The western quoll is now classified as vulnerable in Australia, though the IUCN Red List lists it as Near Threatened, with numbers remaining stable.
  They have been reintroduced to a number of locations, such as the Mt Gibson Wildlife Sanctuary in Western Australia.

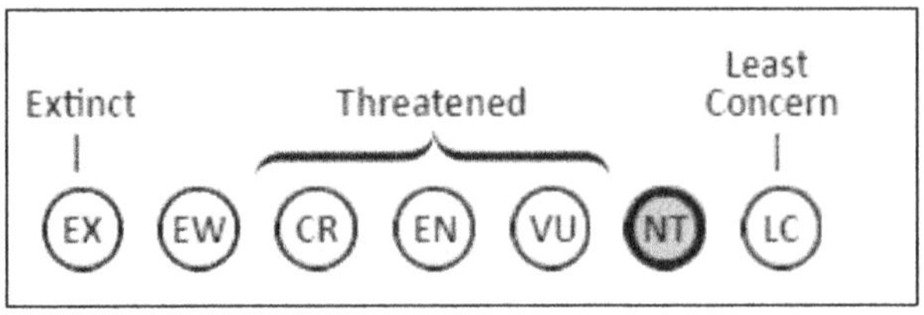

[1]

Little Penguin, Little Blue Penguin, Fairy Penguin, Blue Penguin, Pibaap (Noongar)

Eudyptula minor

Little penguins are the smallest penguin species, standing about 40cm tall and weighing around 1kg. They are known as "fairy penguins" or "little blue penguins" and live in Australia and New Zealand. They can swim at speeds up to 6km/h and are capable of diving to depths of about 60 meters.

- **Physical characteristics and adaptations**
  Little penguins are the smallest of all 17 penguin species. Their blue-grey back and white belly serve as **camouflage** in the water, hiding them from **predators** from both above and below. They have approximately 10,000 feathers that **interlock** to trap warm air and keep them warm and **buoyant**, preventing their skin from getting wet. A special oil from their tail helps keep their feathers waterproof. They use their flippers like wings to swim, reaching speeds of up to 6km/h.
- **Diet and foraging**
  They primarily eat small fish, squid, and krill. They typically dive to depths of 10–30m but have been recorded diving as deep as 72m. They often gather in groups, called "rafts," in the water before heading back to shore, and can spend their days at sea looking for food.
- **Life and breeding**
  Little penguins are often **monogamous** and form strong pair bonds. They dig burrows in soft sand or soil to nest in, breed, and raise their young. Once a year, they go through a **moult** that can last about 15 days. During this time, they cannot swim and must stay in their burrows while their new, waterproof feathers grow in. Chicks are born with a temporary "egg tooth" to help them break out of the egg. After about 8 weeks, they are ready to head to sea for the first time.

- **Other facts**
  They can sleep on the surface of the water, sometimes in short bursts of about four minutes. They make a variety of noises, ranging from a short bark to a growl, to communicate with one another. They are found along the coasts of southern Australia and New Zealand.

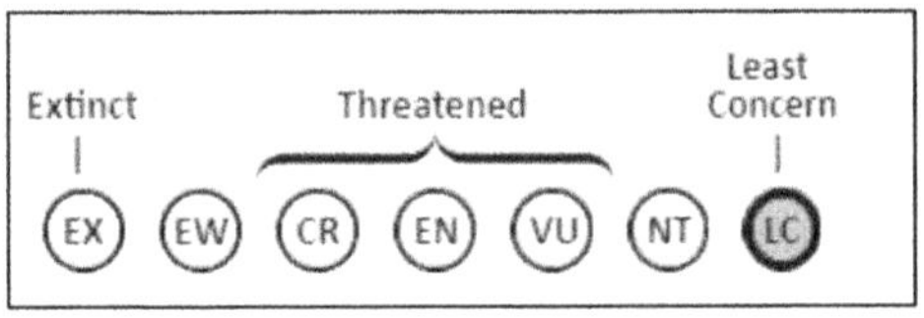

[1]

Black Swan, Maali (Noongar), Marlee or Marli (Noongar)

Cygnus atratus

Black swans are native to Australia, are the only entirely black swans in the world, and are the state emblem of Western Australia. They are large, long-lived, **vegetarian** birds that mate for life, are found in wetlands across Australia, and have been introduced to New Zealand. Some **unique** facts include their ability to get their long necks underwater to feed, their need for a long water runway for take-off, and their temporary inability to fly after breeding season when they must **moult** all their flight feathers at once.

- **Biology and behaviour**
  Black swans are the only entirely black swan species in the world, with broad white wing tips that are only visible in flight. They can live for up to 40 years in the wild. They typically mate for life and are **monogamous**, although they have been known to be unfaithful. Both parents share the duties of raising the young and **incubating** the eggs. They require a long stretch of clear water, at least 40m, for take-off. After breeding, they lose all their flight feathers at once, leaving them unable to fly for about a month. This period of flightlessness coincides with the time the **cygnets** are still dependent on their parents, a convenient arrangement for the swans.
- **Diet and feeding habits**
  They are **vegetarian**, feeding mainly on **algae** and submerged weeds. They feed by "upending" – tipping forward with their long necks and bodies in the water to reach underwater plants, while their feet wave in the air. They can drink and feed in saltwater by collecting the salt in special glands above their eyes and **excreting** it through their nostrils.
- **Habitat and distribution**
  They are found throughout Australia, except for Cape York Peninsula. They prefer to live in fresh, **brackish**, or **saline** water, including large lakes, rivers, and temporary wetlands. They have been introduced to New Zealand, where they are now a common sight on many lakes, and have been introduced to other countries as ornamental birds.
- <u>They are the official state bird of Western Australia</u>.

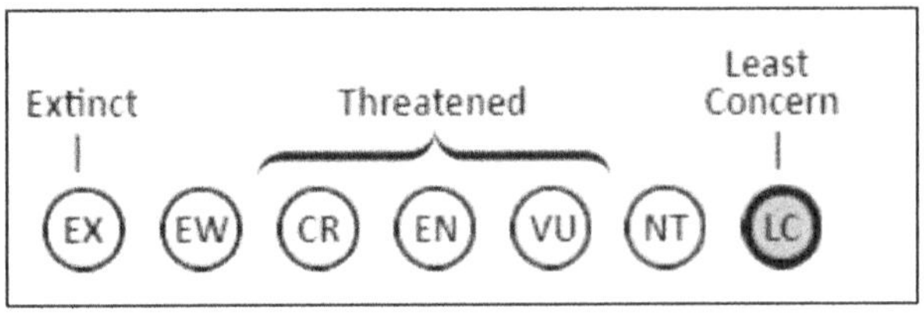

[1]

Honey possum, Noolbenger (Noongar)

Tarsipes rostratus

Honey possums are tiny **marsupials** from southwestern Australia with a very specialised, **nectar**-based diet. Despite their common name, they are not true possums, and they do not eat honey. They weigh a tiny 7-9g when fully grown – a heaped teaspoon of sugar weighs about 7.5g!

- **Unique adaptations and life cycle**
  Honey possums produce the smallest newborns of any **mammal**, weighing only 0.005 grams—about the size of a grain of rice. Male honey possums have the largest **testes** relative to their body size of any **mammal** and the longest **sperm**. Their **testes** can be up to 4.2% of their body weight. To feed on **nectar** and **pollen**, they have a highly specialised, brush-tipped tongue that is nearly a quarter of their head and body length. They can flick this tongue in and out up to three times per second. The honey possum has a very high **metabolic** rate and needs a year-round supply of **nectar** to survive. It can **consume** up to 7ml of **nectar** per day, which is like a human drinking 50L of soda. When food is **scarce** or the weather is cold, honey possums can enter a state of temporary **hibernation** called **torpor** to conserve energy.
- **Pollination powerhouse**
  The honey possum is a "**keystone species**," meaning it has a disproportionately large effect on its **ecosystem** relative to its small size. As the only non-flying **mammal** to rely solely on **nectar** and **pollen** for food, it is a **crucial pollinator** for many flowering plants in its **ecosystem**. Some plants, such as the banksia, have evolved a mutualistic relationship with the honey possum. The possum can discriminate ripe flowers from unripe ones, and its feeding habits are vital to the plant's reproductive cycle.
- **Appearance and behaviour**
  Honey possums are excellent climbers, using their grasping feet and partially **prehensile** tail for balance and grip. This tiny, mouse-like **marsupial** has a long, pointed snout and three dark brown stripes running from its head to its rump. They are primarily **nocturnal** - active at night - and are not particularly **social**. However, several may huddle together to keep warm in cold weather. Honey possums have a very short **lifespan**, living only 1- 3 years in the wild.
- **Habitat and conservation**
  Endemic to Western Australia, they are found exclusively in the flower-rich heathlands and woodlands of southwest Western Australia. Despite being classified as "Least Concern" for **extinction**, their dependence on a constant supply of flowers makes them very **vulnerable** to **habitat** loss from land clearing, wildfires, and climate change. The **unique ecosystems** they **inhabit** are also **vulnerable** to introduced threats, such as the water mould, *Phytophthora cinnamomi*, which causes plant **disease** and can affect their food supply.

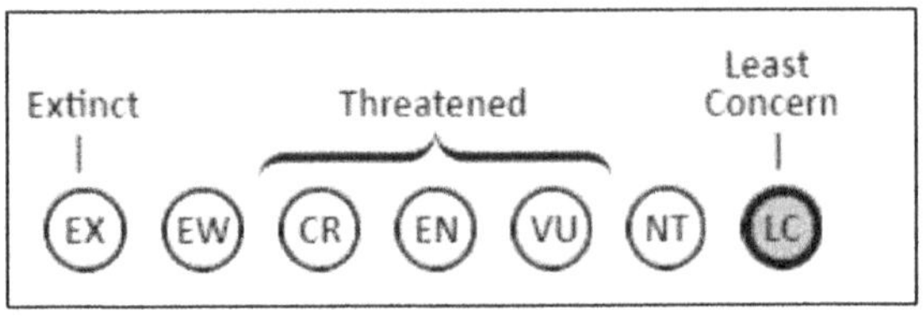

[1]

Tawny Frogmouth, Kambany (Noongar)

Podargus strigoides

- Tawny frogmouths are fascinating Australian birds known for their excellent **camouflage**, which helps them blend in with tree branches during the day. Despite looking like owls, they are actually related to nightjars.
- They are **nocturnal** hunters, using their wide beaks to catch insects and small **prey**, and both parents share the responsibility of raising their young in their loose, stick nests.
- Tawny frogmouths might look like owls with their big yellow eyes, but they are not owls at all! They are more closely related to nightjars, which are also **nocturnal** birds.
- These birds are amazing at hiding. During the day, they sit perfectly still on a tree branch, making themselves look exactly like a broken branch or a dead log with their eyes closed and beak pointing upwards.
- Unlike owls that have sharp **talons** for catching **prey**, Tawny frogmouths have wide, strong beaks that they use to catch bugs, mice, and other small animals; they sit and wait for prey to come within striking distance before catching it with their beaks.
- Tawny frogmouths are excellent parents! The mother lays two or three eggs, and both the male and female take turns **incubating** them and feeding the chicks after they **hatch**.
- Even though they are good parents, they are not great at building nests. They just pile up sticks and leaves, and sometimes the chicks or eggs can even fall out.
- They make a low booming "oom-oom-oom-oom" noise, but if they feel threatened, they can hiss loudly at the same time as opening their beaks and eyes wide, appearing larger and more **intimidating**.
- Tawny frogmouths can survive in many different temperatures, even extreme heat by panting like a dog, and cold by going into a short-term **hibernation** called **torpor**.
- They can be found in **urban** areas, often **roosting** on trees in backyards.
- They have special **asymmetrical** wing feathers for silent flight and longer tail feathers to help them steer while flying.
- Tawny frogmouths stay with the same partner for their entire lives unless one of them dies.

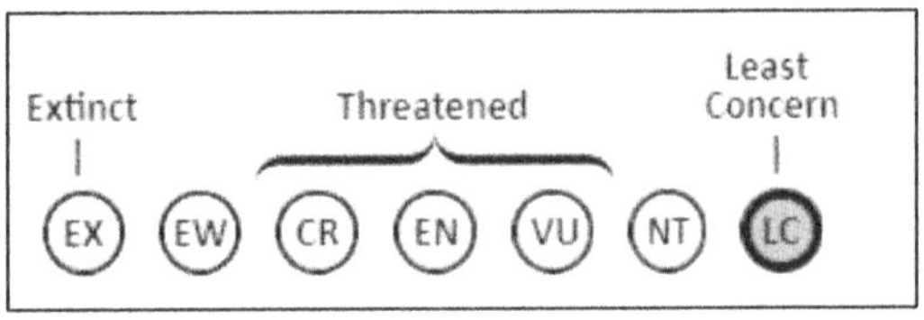

[1]

Bogong moth

Agrotis infusa

The Bogong moth is an **iconic** Australian insect known for its incredible annual **migration**, **social** behaviours, and rich **cultural significance**. For thousands of years, the moths were a **critical** food source for **Indigenous** peoples.

- **Remarkable annual migration**
  Each spring, billions of Bogong moths **migrate** up to 1,000km, flying from their lowland breeding grounds in Queensland, New South Wales, and Victoria to the cooler Australian Alps. They spend the summer **dormant** in alpine caves and rock **crevices** in a state called **aestivation**. Bogong moths are **nocturnal migrants** that use a **sophisticated** internal "compass" to **navigate**. Research has shown that they can **navigate** using the Earth's magnetic field in combination with the night sky. During the summer, the moths pack themselves into caves at incredible **densities**, with up to 17,000 individuals per square metre. This **aggregation** helps them conserve body moisture and regulate their environment. After **aestivating**, the surviving moths return to the lowlands in autumn to mate and lay eggs, completing their life cycle. Their **offspring** will repeat the **migration** the following spring.
- **Key to the alpine ecosystem**
  As they **migrate** to the mountains, the moths create a massive **influx** of high-fat and high-protein food for other alpine animals. The **critically endangered** mountain pygmy-possum relies on the bogong moth as a **crucial** food source after it wakes from its **hibernation**. The crash in moth populations is a significant threat to the possums' survival. By feeding on plants in pasturelands contaminated with **agricultural pesticides**, the moths can **accumulate** trace amounts of **arsenic**. The concentrated waste from generations of moths in their caves has raised soil **arsenic** levels, which may pose a threat to local **predators**.
- **Cultural icon with a storied history**
  For thousands of years, **Indigenous communities** in southeastern Australia gathered at the moths' alpine summer retreats for **ceremonies** and to harvest them. They would roast the fatty moths over fires and eat them whole or grind them into a paste. The annual moth harvests were important **social** events where Aboriginal nations would meet for trade, marriages, initiations, and the exchange of knowledge. Archaeological evidence from Cloggs Cave in Victoria found moth remains on a grinding stone dating back up to 2,000 years, confirming the ancient harvesting practice. The moth's **Indigenous** name, bugung (meaning "brown moth" in the Dhudhuroa language), has given rise to the names of several locations, including Mount Bogong in Victoria.
- **Modern-day challenges**
  On their **migratory** flight path, the moths are strongly attracted to bright city lights, which can disorient them. Famous instances include moths swarming Parliament House in Canberra, located directly on their route. **Droughts** caused by climate change have contributed to a devastating crash in the bogong moth population, leading to the **species** being listed as **endangered** in 2021. The population collapse has severe knock-on effects for the native animals that depend on them.

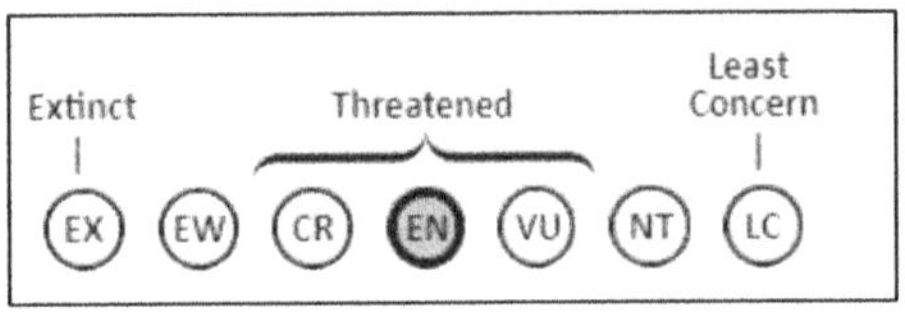

[1]

Red knot

Calidris canutus

- Red knots are shorebirds known for their incredible long-distance **migrations**, travelling thousands of miles between their Arctic breeding grounds and wintering sites in places like Australia and South America.
- There are six **subspecies** of red knots with the Eurasian populations wintering in Africa, Papua New Guinea, Australia and New Zealand.
  Red knots are among the longest-distance **migrants** in the animal kingdom, flying over 14,000km each way between their breeding and non-breeding areas.
- Despite being small birds with a wingspan of only about 47-53cm, they can fly non-stop for very long distances.
- Their diet varies according to season. On the breeding grounds, knots eat mostly spiders, **arthropods**, **larvae**, and some plant material obtained by surface pecking. Red knots love to eat **bivalves** like clams and mussels, which they find by **probing** the mud and sand at low tide. They eat the shells whole, and their strong **gizzards** crush them to digest. They also eat worms and **crustaceans**.
- This species forms enormous **flocks** when not breeding. They often gather in large, tight groups, especially on sandy beaches and mudflats, and can even intermingle with other shorebirds like great knots.
- Red knots change colour! Their feathers turn a rusty red when they are heading north to breed and then return to a more greyish colour for their wintering journey.
- They breed in cold **climes**, including the Arctic tundra, Greenland, the Taymyr Peninsula, the Chukchi Peninsula and some Siberian islands, building shallow nests on the open ground with their eggs **camouflaged** with blotches and spots.
- During their long **migrations**, and in the case of the North American population, places like the Delaware Bay are **vital** rest stops for them, where they feed on horseshoe crab eggs to get ready for the final leg of their journey.
- Red knots are facing challenges due to climate change, which could affect their Arctic breeding grounds and the availability of important food sources.

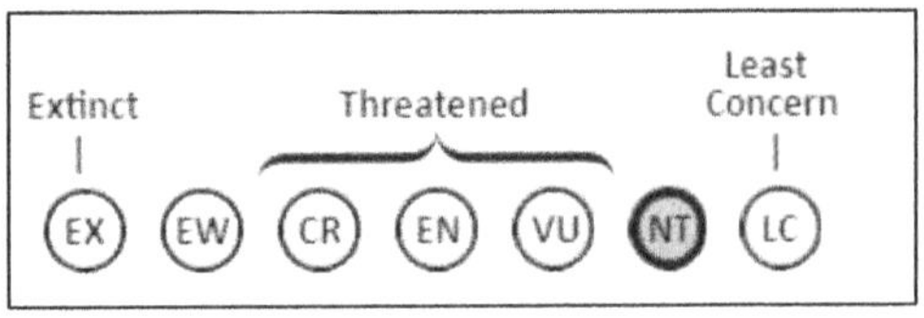

[1]

Red-tailed Tropicbird

Phaethon rubricauda

- The red-tailed tropicbird is the rarest of the tropicbirds and is native to the tropical parts of the Indian and Pacific Oceans. It is similar to a tern in appearance, with almost all-white **plumage**, a black mask, and a red bill.
- The bird has a **streamlined** but solid body shape, a wingspan of 111-119cm, a weight of 800g, and a length between 95-104cm, including the 35cm long red **tail streamers** in adult birds. These **tail streamers** are highly prized by the **Māori people** in the northernmost regions of New Zealand. Still, the birds are seen more frequently in ocean areas with water temperatures ranging from 23- 30°C.
- Common names include red-tailed bos'nbird or silver bos'nbird and strawtail. The New Zealand Māori call it ‘amokura’, and the native Hawaiians ‘koa‘e ‘ula’.
- The red-tailed tropicbird eats mainly squid and flying fish, which it catches by plunge-diving into the ocean or sometimes catching flying fish in the air.
- The red-tailed tropicbird nests in loose **colonies** on **oceanic** islands; the nest itself can be a scrape found on the face of cliffs, on a sandy beach or in a **crevice**.
- Only a single egg is laid and is **incubated** by both sexes for about six weeks. The parents make long trips **foraging** for food during the **incubation** period, sometimes for as much as six days.
- Threats include **habitat** destruction, pest introduction, and **predators** such as rats and **feral** cats that **prey** on eggs and young at nesting sites.
- Red-tailed tropicbirds are generally silent while flying. Birds may greet their mates with a squawk when leaving or arriving at the nest. The chicks chatter repetitively, as a begging call, whenever the parents are near, and the adult birds give a low, growling **defence** call.
- Strong winds can occasionally blow them inland, which explains sightings that have been recorded away from the coastlines and their preferred **habitats**.
- The red-tailed tropicbird is strong in flight but walks with a shuffle on land with some difficulty, landing by flying into the wind, then **stalling** and dropping to the ground. Nests are usually no more than a metre from the edge of the shaded area or scrub, so that it does not have to walk too far.
- They can hover in midair by flying into the wind; pairs may even fly backward and circle each other during courtship. They can be **territorial,** with bill-jabbing and fights breaking out, the two **combatants** locking bills and wrestling for up to 90 minutes.

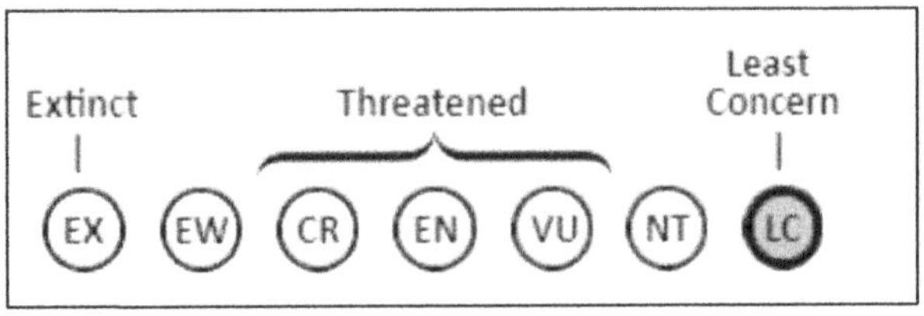

[1]

Bluefin Gurnard

Chelidonichthys kumu

The bluefin gurnard is a bottom-dwelling fish known for its **unique** "walking" fins, noisy **vocalisations**, and striking **pectoral fins** with a bright blue margin.

- **Unique fins and seafloor exploration**
  The first three rays of the bluefin gurnard's **pectoral fins** are separated and modified into finger-like sensory organs. The fish uses these "fingers" to slowly "walk" along the sandy or muddy seafloor. The sensory rays are used to feel and **probe** the seabed, allowing the location of small prey, such as **crustaceans** and worms, buried in the sand. The gurnard's large, fan-like **pectoral fins** are typically olive-green to greyish. When startled or excited, it will flash a vibrant pattern featuring a blue margin and a large black spot with white speckles, possibly to scare off **predators** or attract a mate.
- **Sound production**
  Bluefin gurnards are one of the loudest species in their family and are known to make loud grunting sounds, especially when caught. The sounds are not produced by a voice box but by **specialised** muscles that contract and vibrate against the fish's swim bladder. Research has shown that gurnards can produce different types of sounds. "Grunts" are emitted when in groups, while "growls" are produced at night when the fish are alone.
- **Colour-changing appearance**
  Though often called the "Pacific red gurnard," its natural colouring is a splotchy pale brown. The fish only turns a more vibrant red colour when it is stressed. Its natural brownish colouration helps it **camouflage** with the seabed, where it can bury itself with only its eyes and nostrils exposed to surprise prey.
- **Global and cultural importance**
  The bluefin gurnard is an important **commercial** fish in New Zealand, where it is often caught by bottom-trawling and long-lining. In the Māori language, the fish is known as ‘kumukumu’ or ‘pūwahaiau’. The scientific **species** name, kumu, is a Latinized version of the Māori word ‘koumou’. Despite early settlers in New Zealand sometimes undervaluing it, the gurnard is now a popular food fish with firm, pink, low-fat flesh.

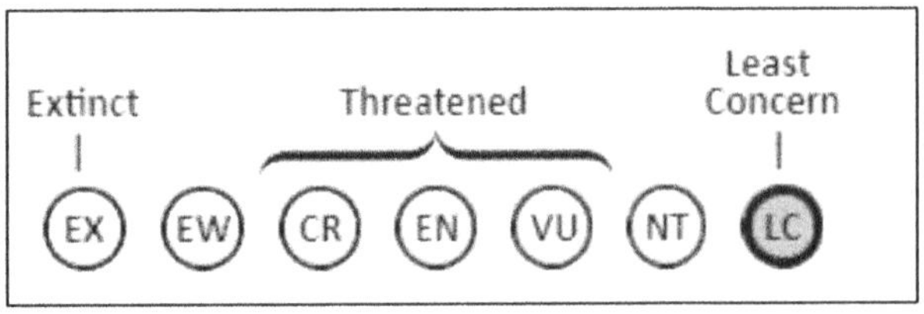

[1]

Curlew Sandpiper

Calidris ferruginea

- The curlew sandpiper is a medium sized wading shorebird with a long black bill, curved downwards at the end, and black feet and legs.
- This **distinctive** beak is what **distinguishes** the curlew sandpiper from other sandpipers of a similar size.
- Although their English name includes the word 'curlew', curlew sandpipers (*Calidris ferruginea*) are not actually curlews!
- Other common names are pygmy curlew, curlew stint and redcrop.
- They are 18-23cm long with a 38-41cm wingspan.
- The curlew sandpiper breeds on the **tundra** of Arctic Siberia. It is a **migratory** shorebird that spends winters mainly in Africa but also in Australia, New Zealand, and southeast Asia, flying thousands of miles. During **migration**, it has been recorded to fly more than 48 hours without stopping.
- During breeding periods, its **plumage** is a rusty red/brown colour.
- Both males and females **incubate** the nests, which usually have 3-4 eggs.
- They are small wading birds, often seen in flocks with other waders.
- They **forage** in soft mud in sheltered coastal areas, mangroves, estuaries, lagoons, and marshes, eating mostly snails, prawns/shrimp, insects, and **polychaete worms**. They can be seen bobbing up and down at the water's edge picking up and probing shallowly for prey.
- Although widespread, **populations** appear to have decreased dramatically in the last **decade** and the IUCN Red List has judged this species as 'near threatened'.
- **Predators** include the Arctic fox which will hunt waders when lemmings are not available or in poor supply.
- Other than **predators**, threats include human impacts such as development, **recreational** activities (4WD, walking, fishing, surfing, boating, etcetera) at **foraging** sites, **groundwater** pollution.

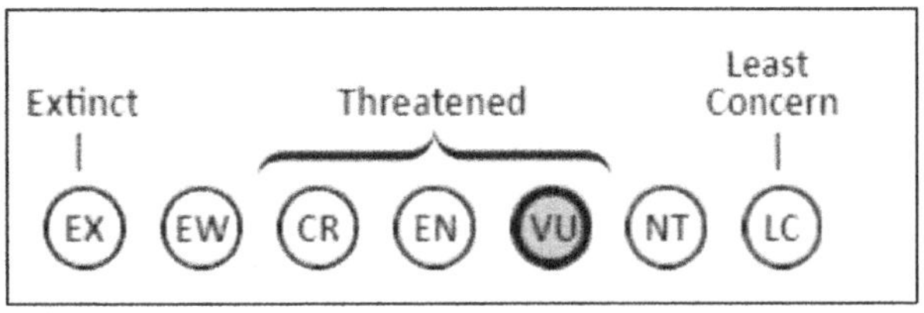

[1]

Feather leaved Banksia

Banksia brownii

The feather-leaved banksia is a **critically endangered** plant species **endemic** to southwest Western Australia. Its name comes from its delicate, feathery leaves, which are among the softest of all *Banksia* species.

- **Conservation**
  Banksia brownii is listed as **Critically Endangered** by the IUCN due to a projected population decline of more than 80% within three generations. The plant is highly **vulnerable** to the introduced soil-borne **pathogen** *Phytophthora cinnamomi* (dieback), which has caused the extinction of multiple populations. Dieback has been so devastating that between 35% and 40% of the species' historical **genetic diversity** has been lost. To save the **species** from **imminent extinction** in the wild, seedlings are being **cultivated** from stored seeds and planted in **disease**-free locations in a process known as **translocation**. Some of this seed came from populations that are now **extinct**, preserving their **unique genetic diversity**. *B. brownii* is the sole **host** plant for the **critically endangered herbivorous** plant-louse *Trioza barrettae*. Conservation efforts have included **translocating** the insect to the new, **disease**-free *banksia* populations to prevent its co-**extinction**.
- **Distinct forms and habitat**
  Originally, two **distinct** forms of were identified, but recent **genetic** testing has confirmed the existence of a third. These forms have different habits, leaf textures, and **inflorescences**. The **species** grows in different **ecological niches** depending on the form and location. Southern coastal populations are found in lower woodland areas on sandy clay soils, often over **laterite** or **granite**.
- **Unique fire ecology**
  The feather-leaved banksia lacks the protective bark or woody **lignotubers** of other banksias, so fire is **fatal** to the mature plant. However, the heat from a bushfire is what triggers the release of the seeds from the woody cones (**serotiny**). For **regeneration** to be successful, fires must not be too frequent, or young plants will be destroyed before they can produce a new seed bank. They must also not be too infrequent, or the seeds will simply die on the plant. The ideal interval is around 18 years.
- **Pollination and reproduction**
  Like many banksias, the plant produces **copious nectar** and is **pollinated** by a range of animals. New Holland Honeyeaters, Western Spinebills, and Red Wattlebirds are frequent, effective **pollinators**. **Nocturnal mammals** like honey possums and bush rats also visit the flowers and carry **pollen**. The species is capable of some **self-pollination**, a rare trait among banksias. This **adaptation** is likely a response to the small, fragmented populations, which can make **cross-pollination** more difficult. Despite its ability to **self-pollinate**, it has a very low fruiting rate, with fewer than 1% of flowers developing into seed-bearing **follicles**. A large portion of the seeds that are produced are then eaten by insects and birds.

- **Appearance**

  The **species**' most **distinctive** feature is its dark green, feather-like leaves, which are divided into up to 70 fine **lobes**. It produces large, **cylindrical** flower spikes, 6 to 19cm long, in a metallic red-brown or golden-brown colour. The spikes appear in autumn and winter. While typically an upright shrub, *Banksia brownii* can range from a low, spreading shrub in exposed areas to a small tree up to 6m tall in more sheltered locations.

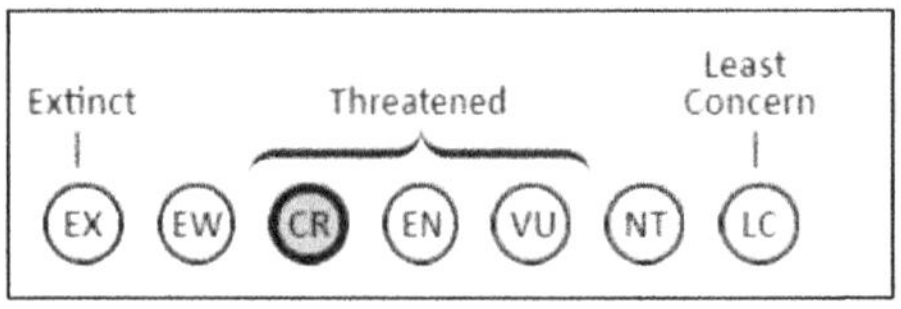

[1]

Broad-faced Potoroo, Moda (Noongar)

Potorous platyops

The broad-faced potoroo was a small, **semi-arid marsupial** from southwestern Australia that became extinct over a century ago. Because few live specimens were ever observed, most facts about this **species** are based on historical accounts and **subfossil** remains.

- **Extinction**
  The broad-faced potoroo had a swift and mysterious disappearance: The first **specimen** was described in 1844, and the last known live **specimen** was captured just three **decades** later in 1875. The **species** was likely **extinct** by 1905. Nyungar peoples from the Margaret River region described the broad-faced potoroo as a slow-moving animal, **vulnerable** to introduced **predators** such as cats. It was also possibly affected by a **disease** outbreak among native **species** in the 1890s. Evidence suggests the potoroo population was already in decline before the invasive European rabbit entered its **habitat**. The rabbits likely worsened the situation by competing for resources.
- **Unique physical traits**
  Its species name *platyops* refers to its wide face. Unlike its longer-nosed cousins, this potoroo had a notably broad skull and a short, blunt muzzle. It was smaller and more **compact** than other potoroos, measuring about 305mm in length with a tail of 178mm. It weighed approximately 800 grams. It had **distinctive** colouring with a coat of grayish-brown on its back and straw-yellow hairs giving it a streaked appearance. Its belly and flanks were a paler, grayish-white colour.
- **Habitat and habits**
  While its exact diet is unknown, it is likely to have had a **specialised** diet. The short, blunt nose suggests it may have eaten underground **fungi**, similar to the **rufous** bettong and the now-**extinct** desert rat-kangaroo.
  Subfossil evidence indicates the **species** once lived across a wide area of **semi-arid** coastal districts in southern and western Australia. This sets it apart from its forest-dwelling relatives, such as the long-nosed potoroo. Very little is known about the animal's behaviour. Early reports mention it was found in **thickets** near salt **lagoons**, and Nyungar informants said it had similar habits to the quokka. After the species was described, additional bones from a cave in South Australia were initially described as a new **species**, *Potorous morgani*. They were later identified as broad-faced potoroo remains.
- **A scientific enigma**
  The **species** is a powerful reminder of how little is known about many **species** that went **extinct** shortly after European **colonisation**, making it difficult to understand or reverse their decline fully.

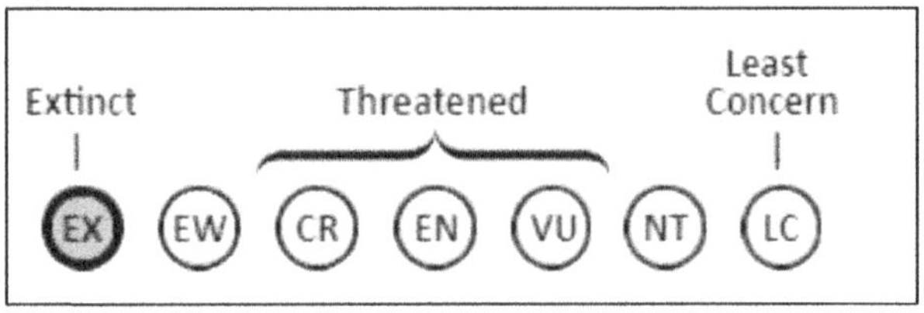

[1]

Sundew species

Drosera oreopodion

Whilst there are numerous species of sundew which occur in Noongar country, this particular one is classified as Critically Endangered on the IUCN Red List.

This species is a pygmy sundew known for its diminutive size, unusual flowering stem, and critically endangered status. Like all sundews, it is a **carnivorous** plant that uses sticky glands on its leaves to capture, **digest**, and absorb **nutrients** from insects.

- **Key facts**
  It is one of the smallest **carnivorous** plants: Its leaves are only about 5.5mm long, with a circular, sticky **lamina** just 1.5mm wide.
  Despite its miniature rosette, the plant's flowering stem (**inflorescence**) is disproportionately tall, growing to 3.5cm - this keeps **pollinators** from getting caught in the plant's traps.
- **Habitat and discovery**
  This rare **species** is known from only a few hundred plants in a very small **habitat** area in Western Australia. It is found in the foothills of the Darling Range near Perth, where it grows in seasonally moist, clayey sand. The plant was unknown to science until it was discovered by **carnivorous** plant expert Allen Lowrie in 1987.
- **Unique characteristics**
  As a pygmy sundew, it reproduces **asexually** by producing **specialised** scale-like buds called **gemmae**, typically during the autumn. It has highly branched, blackish-red style arms that resemble the tentacles of a sea anemone, which help distinguish it from other sundews.
  Some botanists suggest that *Drosera oreopodion* may be of hybrid origin, possibly from *Drosera micrantha* and *Drosera helodes,* given its intermediate appearance and the fact that some flowers show faint pink blushing reminiscent of the latter **species**. However, it is fully capable of producing viable seeds.

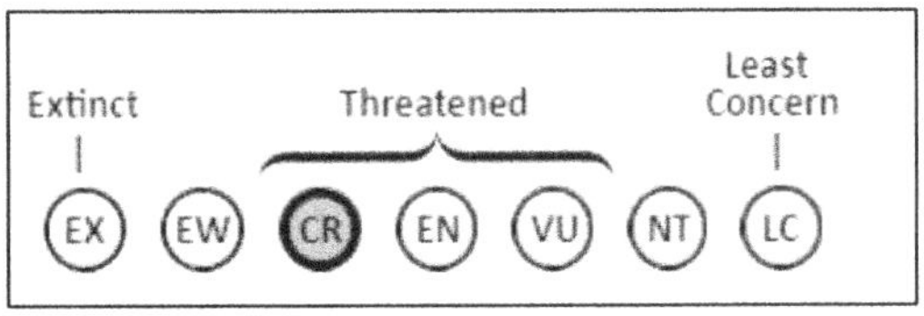

[1]

Brown Argonaut, Winged Argonaut, Brown Paper Nautilus, Lesser Argonaut

Argonauta hians

The brown argonaut is a rare, **pelagic** octopus known for its extreme sexual **dimorphism**, unusual **reproduction**, and shell-like egg case.

- **Unique biology of the female argonaut**
  Unlike the hard, protective shells of true nautiluses, the brown argonaut's paper-thin, coiled case is an egg case. Female argonauts secrete this delicate case from **specialised glands** in two of their modified arms and use it as a **brood chamber** for their eggs.
  The female argonaut uses her egg case to control her **buoyancy**. She can regulate the air and water pressure inside the case, allowing her to float effortlessly through the open ocean without expending much energy. These clever octopuses have been observed hitching rides on jellyfish and other **organisms**. They use their **host** as a hunting platform, and even as a way to hide from **predators**.
- **Radical size difference between sexes**
  Male brown argonauts are incredibly tiny, measuring less than 3cm long, while females can grow up to 27cm, with their egg case sometimes reaching 30cm. This means a female can be up to 600 times heavier than her male counterpart. Due to their small size, male argonauts were not discovered until the 19th century, long after females had been documented.
- **Strange reproduction**
  For mating, the male uses a modified tentacle called a **hectocotylus**, which is stored in a sac under his eye. During mating, this arm, laden with **sperm**, detaches and crawls into the female's **mantle cavity**. The male is believed to die soon after. The female stores the detached male arm and **sperm** in her **gill cavity** and can mate with multiple males. She can then **fertilise** her eggs as she produces them, sometimes months later.
- **Ancient and mysterious history**
  For centuries, scientists believed that argonauts were like hermit crabs that stole their "shells" from other creatures. It was only in the 19th century that marine biologist Jeanne Villepreux-Power demonstrated that argonauts construct their own egg cases. One theory suggests that argonauts may have originally used discarded ammonite shells for their eggs before developing the ability to create their own.
- **Fossils from the Pliocene**
  The oldest known fossils of the brown argonaut are from the middle Pliocene Epoch, millions of years ago.

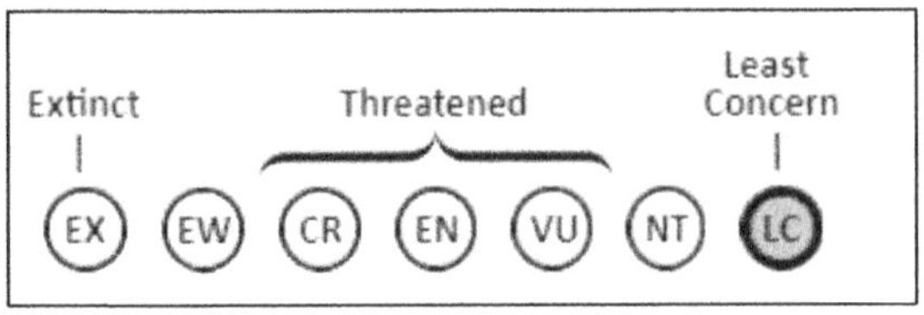

[1]

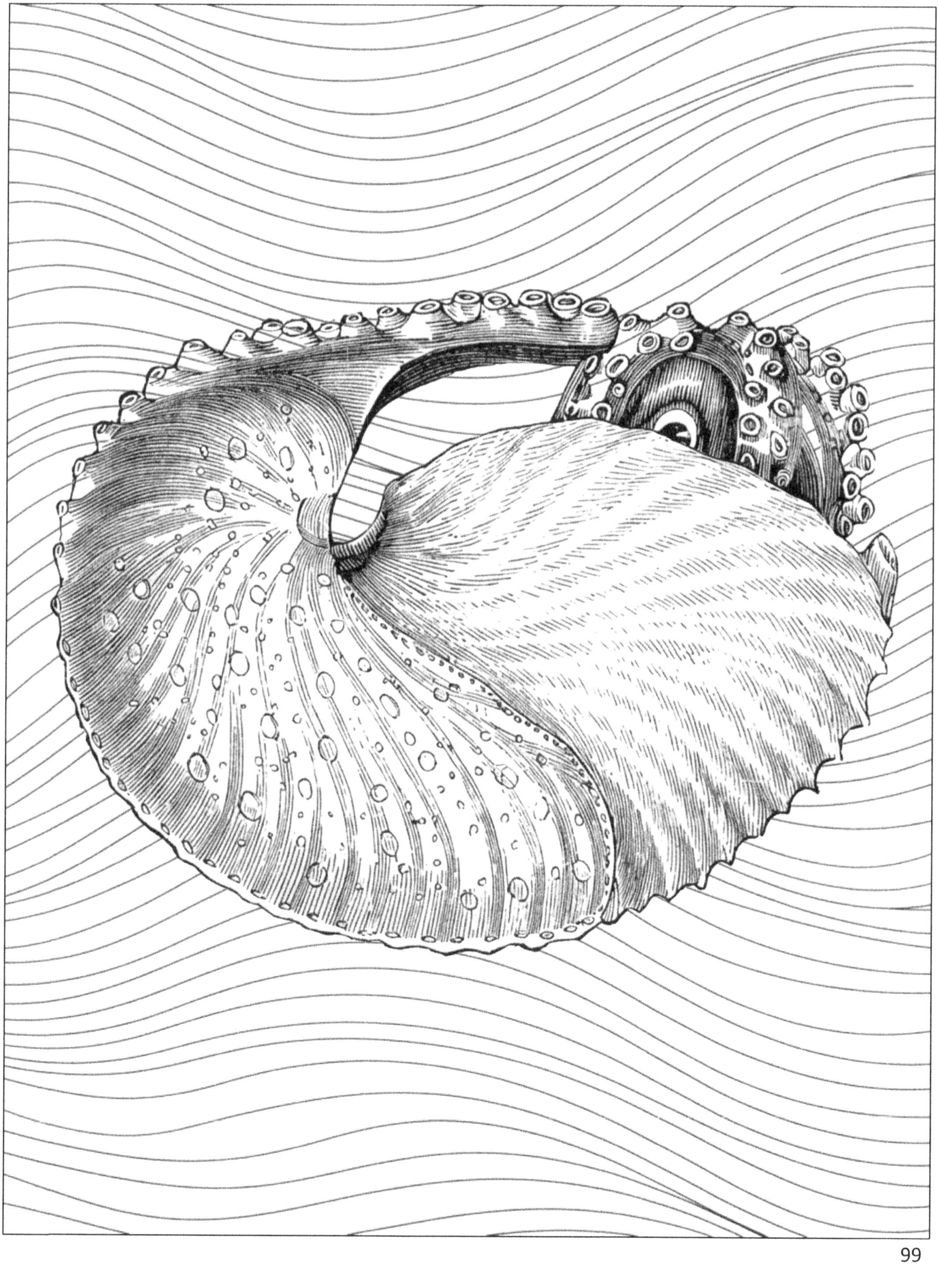

Australian Shelduck, Nyimarak (Noongar), Guraga (Noongar)

Tadorna tadornoides

- The Australian shelduck, also known as the chestnut-breasted shelduck or mountain duck, is a large, distinctive waterfowl species found in southern Australia.
- They **inhabit** freshwater and some saltwater **habitats** near freshwater sources and are often seen in pairs or large **flocks**.
- <u>They are among the largest duck species in Australia</u>.
- Shelducks are a group of large goose-like ducks , part of the bird family *Anatidae*. The genus name *Tadorna* comes from Celtic roots and means “pied waterfowl”.
- They are known for their striking **plumage**, including a chestnut breast and black body with white markings. Males have a black head, females have white around the eye and **bill**, a white neck ring, and a chestnut breast.
- Australian Shelducks stand upright when grazing on land and also sit high in the water.
- Males produce a deep, harsh honking sound, while females make a high-pitched “ank-ank” sound.
- They nest in tree hollows, holes in banks, or even rabbit burrows, often lined with down.
- Chicks leave the nest within a couple of days, led by their parents, and often form **crèches** with other families.
- Australian Shelducks are **omnivores**, feeding by grazing, dabbling and filtering. They eat **tubers** and other plant material, such as saltbush and samphire, along with a range of insects, **crustaceans** and molluscs.
- They have adapted to various **habitats**, including coastal areas, freshwater swamps, and even inland lakes and gravel pits.
- After breeding, they may **migrate** to wetlands in large **flocks** of 1000 or more to **moult**
- their feathers.
- They are generally **monogamous**, with both parents participating in raising the young.
- While not a primary game bird, they may be hunted occasionally due to their habit of grazing on crops.

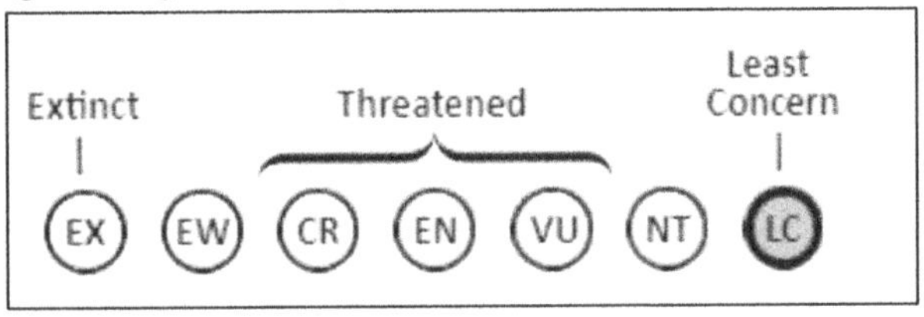

[1]

EPBC List = Least concern

Slender Seamoth, Long-tailed Dragonfish, Pelagic Dragon-fish, Winged Dragonfish, Longtail Seamoth

Pegasus volitans

Whilst this species is listed as Data Deficient on the IUCN Red List, it is just such a weird-looking fish that I could not help but include it in this book!

Slender seamoths are interesting for their armour-like bony plates, their ability to "walk" on the seafloor using modified **pelvic fins**, and their excellent **camouflage**, which they can change rapidly to match their surroundings. Unlike seahorses, they lack swim bladders and lay eggs in open water, relying on a **unique**, one-piece skin-shedding method to remove parasites. They are also believed to form **monogamous** pairs.

- **Physical and behavioural facts**
  Seamoths are encased in fused, bony plates and have a flattened body. They use their finger-like pelvic fins to move along the bottom, giving the appearance of walking. They have a strong ability to change colour to match their **environment**, a primary **defence mechanism** against **predators**. They shed their skin in one piece, often with a rapid jump, to get rid of **parasites** or **algae**.
- They do not **brood** their young like seahorses; instead, they **spawn** in open water.
- It is believed that seamoths form **monogamous** pair bonds.
- They have large, wing-like pectoral fins that they often hold out **horizontally**.
- **Habitat and diet**
  Seamoths are bottom-dwellers found on soft, sandy or muddy **substrates** in the Indo-Pacific region, from shallow coastal waters to about 90m deep.
- They use their long, probing snouts to hunt small **benthic invertebrates**, such as tiny **crustaceans** and worms.
- **Additional facts**
  Their small mouth is located under the snout and can be protruded to catch prey. Adults display various colours and patterns to **camouflage**, while **juveniles** are sometimes entirely black. Males can have a greenish margin on their **pectoral fins** for display.
- The slender seamoth can reach about 18cm in length.

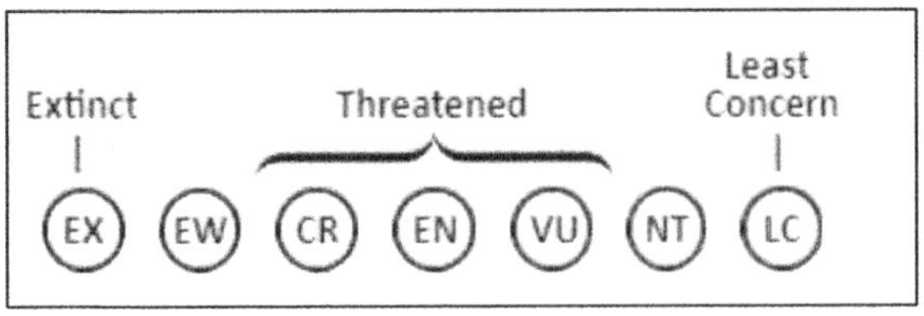

[1]

White-bellied Sea Eagle, Djilyin/Djilyiyan/Ngoolor (Noongar)

Haliaeetus leucogaster

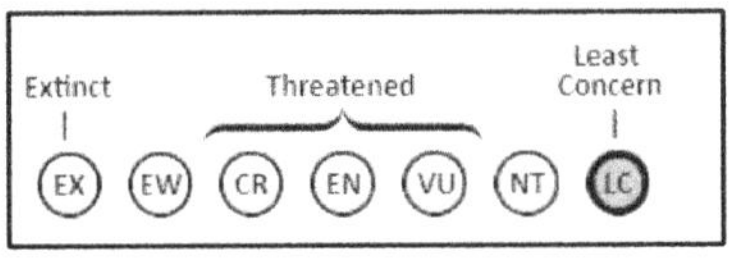

[1]

- The white-bellied sea eagle is a large, **diurnal raptor** with long, broad wings and a wedge-shaped tail. It measures 75–85cm in length and has a 180–220cm wingspan. Females weigh between 2.8kg and 4.2kg and are larger than the males, which weigh between 2.5kg and 3.7kg.
- The **plumage** of adult birds is mainly white and grey. The head, breast, and belly, and the feathering on the legs are white. The back and upper surfaces of the wings are grey, although the wings have black tips. The undersides of the wings are greyish-black around the **distal** edges, with a smaller area of white along the leading edge. The tail is grey at the base and has a white tip. The bill is bluish-grey with a blackish tip, the iris is dark brown, and the legs and feet are a cream colour. The **juveniles** differ from the adults in appearance, and there is a gradual transition from the brown and cream **plumage** of **juvenile** birds to the white and grey **plumage** of the adults across a series of **moults** over a period of several years.
- The white-bellied sea eagle is generally seen singly or in pairs, though it may occasionally **congregate** around sites where food is **abundant**. Its call is a loud, goose-like honking.
- The white-bellied sea eagle breeds and hunts near water, and fish form around half of its diet. It **opportunistically consumes carrion** and a wide variety of animals. They are skilled hunters that can attack **prey** as large as a swan. They also eat **carrion** like fish and sheep.
- **Human disturbance** to its **habitat** is the main threat, both from direct human activity near nests and the removal of suitable trees for nesting. The white-bellied sea eagle is **revered** by **indigenous people** in many parts of Australia and is the subject of various folk tales throughout its **range**.

Osprey, Yoon-door-doo(Noongar)

Pandion haliaetus

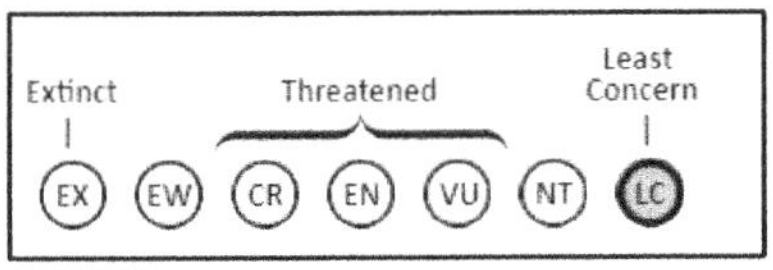

[1]

- An osprey may log more than 160,000 **migration** miles during its 15-to-20-year lifetime. Scientists track ospreys by strapping lightweight **satellite transmitters** to the birds' backs. The oldest known osprey was at least 25 years, 2 months old.
- Ospreys are unusual among hawks in possessing a **reversible** outer toe that allows them to grasp with two toes in front and two behind. **Barbed** pads on the soles of the birds' feet help them grip slippery fish. When flying with **prey**, an osprey lines up its catch head-first for less wind resistance.
- Ospreys are excellent **anglers**. Over several studies, the average time they spent hunting before making a catch was about 12 minutes.... <u>something to think about next time you throw your line in the water</u>!
- The osprey readily builds its nest on man-made structures, such as telephone poles, channel markers, **duck blinds**, and nest platforms designed especially for it. Such platforms have become an important tool in re-establishing ospreys in areas where they had disappeared. In some areas, nests are placed almost exclusively on **artificial** structures.
- Osprey eggs do not hatch all at once. Rather, the first chick emerges up to five days before the last one. The older **hatchling dominates** its younger **siblings** and can **monopolise** the food the parents bring. If food is abundant, chicks share meals in relative harmony; younger ones may starve to death in times of **scarcity**.
- Their underwater eyesight is a superpower. Ospreys have a vision that is well adapted to detecting underwater objects from the air. **Prey** is first sighted when the eastern osprey is 10- 40m above the water, after which the bird hovers momentarily and then plunges, feet first with wings raised; a large splash is made as it enters the water.
- Ospreys are found near open water with an **abundant** supply of fish; listen for the osprey's whistling or chirping calls overhead, or look for this bird's **distinctive** flight profile and heavy wingbeats.

Baudin's Black-Cockatoo, Long-Billed Black-Cockatoo, Ngolak (Noongar)

Zanda baudinii

Baudin's black cockatoos are known for their long, narrow beaks, which they use to feed on seeds, **larvae**, and fruits from native forests, and they are often **distinguished** by the pink eye-ring on males and a grey one on females. These large parrots are highly social, nesting in hollow trees and living for up to 50 years, but they face **severe** threats from **habitat** loss and illegal shooting and are listed as endangered.

Physical characteristics

A key feature is their long, narrow upper bill, which **distinguishes** them from the similar Carnaby's cockatoo. Their black feathers are edged with white, giving them a scalloped look. They have a distinct white panel on their tail feathers. Males have a pink eye-ring and a black **bill**, while females have a grey eye-ring and a grayish-white **bill**. They have a short crest of longer black feathers on their head that can be raised or lowered.

Behaviour and diet

They are **social** birds that are often seen in pairs or small groups. They feed on seeds from eucalyptus, banksia, and hakea, as well as fruit from apples, pears, and other trees. They strip the bark off dead trees to find beetle **larvae**. They nest in hollow trees, using the same hollow year after year.

Conservation status and threats

Baudin's black cockatoo is listed as endangered under both Western Australian and federal legislation. Their populations have declined significantly due to land clearing for **agriculture** and other developments. A major threat is illegal shooting by orchardists who consider them pests because of their fondness for fruit. They face competition for nesting hollows from **feral** European honeybees, which can be a severe problem.

Life cycle

They can live for more than 50 years in the wild. They begin to breed at about 4 to 5 years of age. They lay two eggs each year, usually in October.

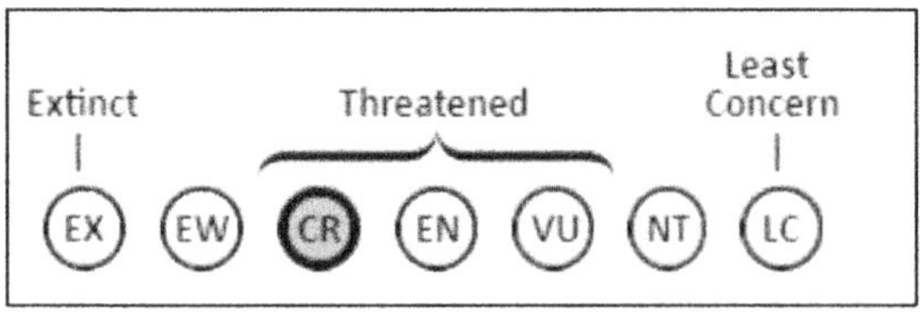

[1]

Sooty Albatross

Phoebetria fusca

- Sooty albatrosses are seabirds with a striking appearance and interesting behaviours.
- The sooty albatross is also known to sailors as the Quaker,
- They have a dark, sooty-brown to black plumage, a white ring around their eye, and a black beak with a yellow or orange stripe.
- These birds are known for their impressive wingspan, which allows them to glide effortlessly over the ocean for long distances.
- Sooty albatrosses are medium-sized albatrosses, with a wingspan of about 2m.
- Their **plumage** is a dark, sooty-brown to black colour. The head and sometimes the tail can be a darker shade than the rest of the body. They have a **distinctive** white ring around their eye and a yellow or orange stripe on their beak.
- In general, the sooty albatross can be found **foraging** over the deep ocean. Non-breeding adults do not often venture out of their native oceans but have a broader **foraging** range. In contrast, breeding adults **forage** closer to their breeding sites (within 600km) and **juvenile** birds tend to be confined to **subtropical** waters.
- They eat fish, **crustaceans**, squid, and **carrion**. They often **forage** alone, but may join mixed-species **flocks** at times, especially when following fishing boats.
- Sooty albatrosses are masters of the wind. They use their long wingspan to glide for long distances, often without flapping their wings, making them very efficient flyers.
- They breed in loose **colonies** on cliffs or steep slopes. A single egg is laid in October or November, and both parents help to **incubate** it and raise the chick.
- Like all albatrosses, sooty albatrosses drink saltwater. They have special glands above their eyes that help them get rid of the excess salt, which can sometimes drip from their nostrils.
- Sooty albatrosses can live for a long time, with some individuals living for over 60 years.
- Sooty albatrosses are currently listed as near threatened due to threats like bycatch in fishing gear and climate change.

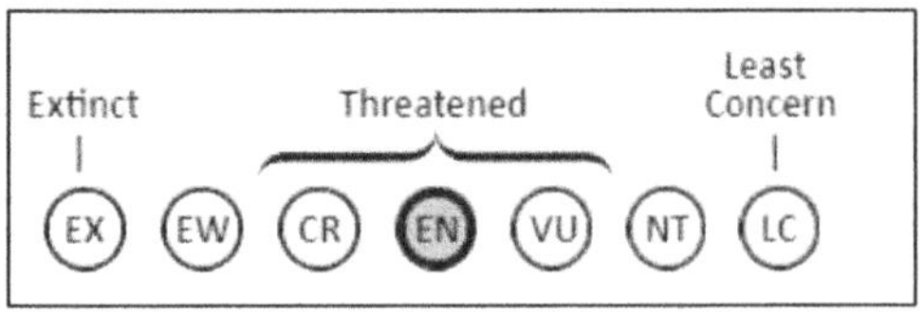

[1]

Southern Emu-wren

Stipiturus malachurus

The Southern Emu-wren found in southwestern WA has an exceptionally long, **filamentous** tail, is often shy, and prefers to run like a mouse in dense, low-lying vegetation like coastal heathlands and sandhill scrub. A blue chin and throat distinguish males, while females lack this colouration. They feed on small insects, build dome-shaped nests low to the ground, and are often heard before seen because of their soft calls.

- **Appearance**
  The tail is about twice the length of its body and resembles an emu's feathers. Males have a blue throat, upper chest, and eyebrow, along with a streaky grey-brown back. Females are similar but lack the blue colouring. Despite the long tail, the body is very small, making the entire bird only a few centimetres long. These tiny birds weigh about the same as a teaspoon of sugar (5g) and their eggs are the size of a pea!
- **Behaviour and habitat**
  They are ground-dwellers, often running like a mouse with their tail held upright as they **forage** for small insects in **dense vegetation**. They **inhabit** moist, **dense** scrublands, coastal heathlands, and sandhill **vegetation**. They are weak flyers and prefer to stay low in the cover of **vegetation**. You may hear their soft, high-pitched calls before spotting them.
- **Breeding and diet**
  Their diet consists almost exclusively of small insects. They build dome-shaped nests low to the ground in **dense vegetation**, weaving them with grasses and fine twigs. Breeding occurs between August and January. Their eggs are approximately the size of a pea! Both parents feed the chicks, which become independent within about three months.

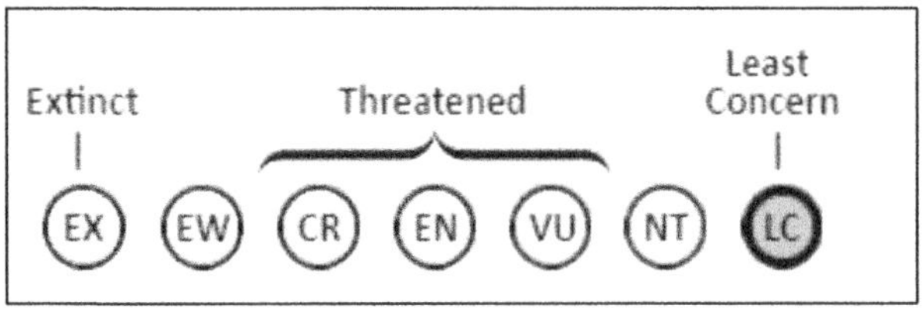

[1]

Spiny Tigertail, Tigertail Dragonfly

Archaeosynthemis spiniger

The spiny tigertail is a vulnerable dragonfly known for its black and yellow markings and its **habitat** in streams and rivers in south-western Australia. Interesting facts include its ancient family lineage, the **larvae**'s ability to survive drought by burying themselves in mud, and its **specific habitat** needs for breeding.

- Distinctive markings: It is a medium-sized dragonfly with **prominent** black and yellow colours.
- Family and habitat: It belongs to the *Synthemistidae* family, which is known for both sluggish streams and fast-flowing rivers, and the spiny tigertail inhabits streams and rivers in south-western Australia.
- **Larval adaptation**: The **larvae** of some *Synthemistidae* **species** can survive **drought** by burrowing into mud or sand. The female spiny tigertail lays eggs in the mud, and they may not hatch until there is enough water.

[1]

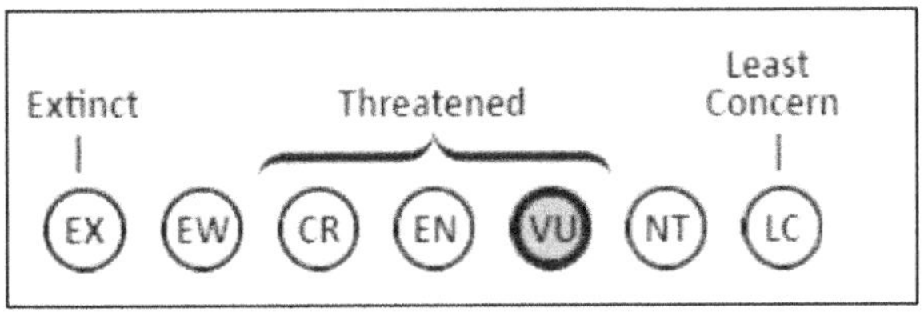

So, what is a dragonfly?

- Dragonflies are flying insects, easily recognised by their long bodies with two pairs of strong, transparent wings. There are over 5,000 different species of dragonflies; in fact, the largest insect known to inhabit Earth was a dragonfly called the *Meganeuropsis permiana*. This huge insect had a body length of almost 43cm. Nowadays, we see much smaller dragonfly species, averaging around 2.5-10.5cm in length.
- Dragonflies have nearly 360-degree vision.
- Dragonflies can fly backward.
- If a dragonfly lands on you, it is considered by many people to be a sign of good luck!
- Just one dragonfly can **consume** over one hundred mosquitoes in a day.
- In ancient times, people believed dragonflies were actual dragons due to their impressive flying abilities and **vivid** colouring.
- Dragonflies use the sun to help warm their body, by extending their wings to **absorb** as much heat as they possibly can.
- Dragonflies typically eat other insects, such as midges, mosquitos, moths, bees, and flies. Their **nymphs** tend to eat tadpoles and other small creatures that they encounter while living underwater.
- Dragonflies are excellent flyers - reaching speeds above 30mph, which helps them to hunt successfully. They also have very strong jaws and sharp teeth which helps them to catch and eat other insects.

Brush-tailed Bettong, Woylie (Noongar)

Bettongia penicillata

Brush-tailed bettongs, also known as woylies, are **crucial** "**ecosystem** engineers" that play a **vital** role in soil health and seed **dispersal**. They are **nocturnal marsupials** that use their **prehensile** tails to carry nesting materials, can breed year-round, and have a **unique** diet of **fungi, tubers**, and seeds. Woylies can dig between 2 and 6 tonnes of dirt annually, which improves water **infiltration** and **nutrient** cycling in the soil.

- **Ecosystem role**
  They turn over topsoil and cycle **nutrients**, improving soil aeration and water **infiltration**. They spread the seeds of native plants, which helps with **regeneration** and plant recruitment.
- **Lifestyle and behaviour**
  They are most active from dusk until a few hours before dawn. They build dome-shaped nests out of grass and bark, often in a depression scraped under a bush. They use scent, including urine and scent glands, to communicate. They use their long, **prehensile** tails to carry nesting materials like grass and sticks.
- **Diet**
  Their diet is primarily composed of underground **fungi**, which they locate with a keen sense of smell. They also eat **tubers**, bulbs, seeds, insects, and resin. They get most of their moisture from their food and have a very low need for drinking water.
- **Reproduction**
  They can breed continuously throughout the year. Females can become sexually **mature** by about six months of age. After giving birth, a female can mate again almost immediately. A **joey** stays in the mother's pouch for about 90–110 days.

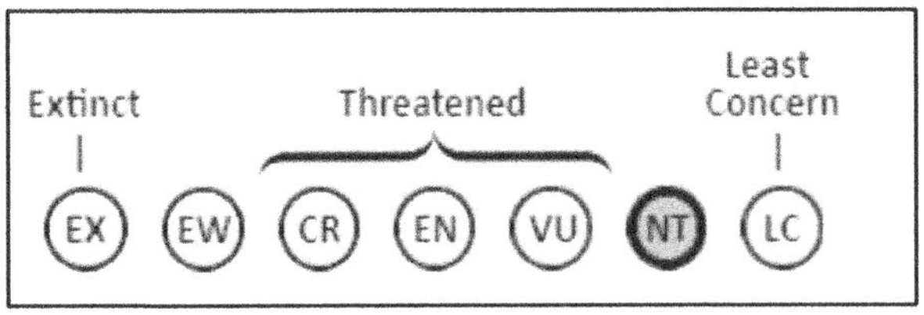

[1]

Marbled Gecko, Marbled Southern Gecko, Bibdjool = gecko (Noongar)

Christinus marmoratus

Marbled geckos are **social**, **nocturnal** lizards that can live in groups of up to 10 individuals. They are famous for their ability to drop their tails when threatened and their thick tails, which store fat for **hibernation**. They are also notable for their excellent climbing skills and their beautiful, marbled-like skin patterns.

- **Social and behavioural facts**
  Unlike many **territorial** geckos, marbled geckos are often found living in groups of up to 10, possibly for mating, **defence**, or to share good hiding spots. They are active at night, hunting insects, and are often found near outdoor lights where insects gather. They regulate their body temperature by changing their posture, such as flattening their bodies on warm rocks or lifting them to cool down.
- **Physical and anatomical facts**
  They can drop their tails when threatened to aid in escape. The thick tails store fat for energy during **hibernation**. If a tail is dropped, it takes about eight months to **regenerate**. The **regenerated** tail has a distinct change in colour and pattern where the break occurred. Their large toe pads are covered in millions of microscopic, hair-like structures called **setae**. These **setae** create a strong **adhesive** force (known as **Van der Waals' force**) that allows them to climb smooth, **vertical** surfaces like windows and walls. Their name comes from their **variable** colouration, which can range from tan to brown, grey, or reddish, and features patterns that can look like marbling. This provides excellent **camouflage** against rock and tree bark, making them hard for **predators** to spot during the day.
- <u>Marbled geckos lack eyelids and instead use their tongue to lick their eyes clean and moist</u>.
- **Habitat and distribution facts**
  They can be found in rocky areas, under tree bark, in fallen logs, and in man-made structures like buildings and walls.
  Marbled geckos are native to Southern mainland Australia, Tasmania, and parts of Victoria. <u>They are the most southerly distributed gecko in Australia</u>.

Fairy Tern

Sternula nereis

Fairy terns are small, white birds with a black patch near the eye, known for nesting in simple scrapes on sandy beaches and feeding on small fish, often swallowing them head-first. They are a **vulnerable species** in Australia and critically endangered in New Zealand, facing threats from predation and habitat disturbance from human activities. Both parents share the **incubation** and chick-rearing duties, which include teaching their young to catch fish.
The Fairy tern is the smallest of all tern species that breed in southwestern Australia.

- **Behaviour and life cycle**
  Fairy terns make their nests in a simple scrape on the sand, often decorated with shells or pebbles. This is typically above the high-tide line, sometimes with a few small shells or seaweed for lining. They form **monogamous** breeding pairs and raise one or two chicks per season. Both parents share the responsibilities of **incubating** the eggs and raising the young. They often nest in large **colonies** with other fairy terns. The Fairy tern's breeding season in southwestern Australia occurs from October to February.
- **Diet**
  Fairy terns primarily eat small fish, which they catch by diving headfirst into the water. They may also eat small **crustaceans** and **molluscs**. Fairy terns are expert plunge-divers, hovering above the water before diving headfirst to catch small baitfish like whitebait and hardyheads. They typically hunt in shallow water, often alongside larger **predatory** fish.
- **Habitat**
  Fairy terns are found along the coastlines and on islands in Australia, New Zealand, and New Caledonia. Their preferred **habitats** include sheltered sandy beaches, spits, and islands, as well as **estuaries** and **lagoons**. The eggs and young chicks are well **camouflaged** to match the sand and shell grit of their nests. This makes them hard to spot, but also at high risk of being accidentally trampled by beachgoers.
- **Conservation status and threats**
  Their eggs and chicks are **preyed** upon by introduced **mammalian predators** like foxes, stoats, and cats, as well as native **predators**. Human activity, such as people and dogs on beaches, can cause the adults to abandon their nests, leaving eggs or chicks **vulnerable** to **predators** and harsh conditions. Land clearing and development have reduced their breeding and **foraging habitats**.

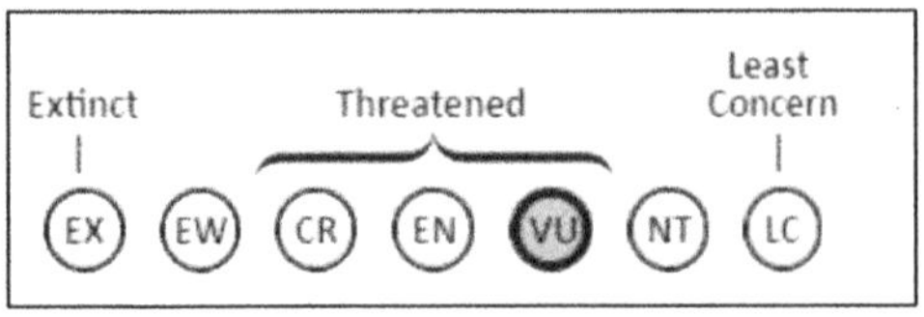

[1]

Common Seadragon, Lucas' Seadragon, Weedy Seadragon

Phyllopteryx taeniolatus

The common seadragon is a marine fish native to Australia that uses leaf-like **appendages** for **camouflage** and has a **unique** breeding cycle where the male carries the eggs on his tail. These fish are poor swimmers and feed by sucking small crustaceans through a tube-like snout. The species is legally protected in some areas due to threats from **pollution** and collection.

- **Camouflage and movement**
  Leaf-like **appendages** and **cryptic** colouration allow the common seadragon to blend in with seaweed and seagrass beds, helping it hide from **predators** and sneak up on prey. They are very poor swimmers, which is why they are sometimes found stranded on beaches.
- **Diet**
  They feed on tiny **crustaceans** like **mysid shrimp**, sea lice, and **larval** fish. They use their long, tube-like snout to suck prey into their mouth. Common seadragons do not have teeth or a stomach, so they must eat constantly.
- **Reproduction**
  Breeding occurs seasonally, from June to January. The male carries the developing eggs on the underside of his tail for about 30-38 days until they hatch. A single **brood** can have up to 250 eggs. The young are free-swimming upon hatching.
- **Other facts**
  Males have narrower bodies and are a darker colour than females. Their colour can vary depending on their **habitat** and food source. The species is legally protected in some regions, such as New South Wales and Tasmania.
- They are the marine state faunal emblem for Victoria.

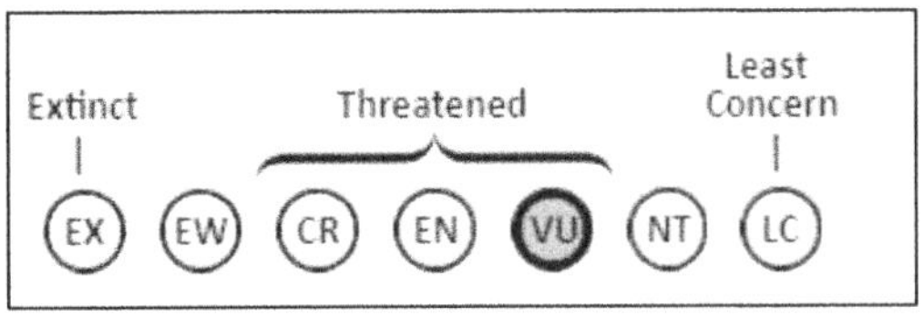

[1]

Bottlenose Wedgefish

Rhynchobatus australiae

The bottlenose wedgefish is a ray with a "shark-like" appearance. Also known as the "rhino ray," this **unique species** is among the most **imperilled** marine fish in the world, primarily due to fishing pressure and **habitat** loss.

- **Unique biology and appearance**
  Though commonly mistaken for a shark due to its large **dorsal** and tail fins, the bottlenose wedgefish is actually a type of ray. Its body is flattened on the underside, with **gills** located on its belly, but its head and fins have a distinct shark-like silhouette. The nickname "rhino ray" comes from its long, pointed, and bottle-shaped snout. **Juveniles** are grey or brownish with scattered white spots and a black spot above each **pectoral fin**. Adults become much darker, appearing almost black from above, and have three distinct white spots over the back of their **pectoral fins**. This species can grow up to 3m long and weigh as much as 227kg. It has a relatively long **lifespan**, with females living up to 47 years and males up to 40 years.
- **Habitat and nocturnal behaviour**
  The bottlenose wedgefish is a **demersal species**, meaning it lives and feeds on or near the seabed. It prefers shallow coastal waters with soft, sandy bottoms and is often found near coral reefs, at depths up to 60 meters. Studies using **acoustic** tracking have revealed that the wedgefish is primarily **nocturnal**, with its **nocturnal** activity becoming more pronounced in warmer summer temperatures. When resting on the seafloor, it uses **spiracles** behind its eyes to actively pump water over its **gills** for breathing. Its diet consists of **crustaceans**, **molluscs**, and other bottom-dwelling fish. It has powerful plate-like jaws adapted for crushing hard-shelled prey.
- **Conservation challenges**
  The International Union for Conservation of Nature (IUCN) has assessed the bottlenose wedgefish as Critically Endangered, with populations estimated to have declined by over 80%. The wedgefish's large, valuable fins are highly prized in the international shark fin trade, making it a target for fishing and bycatch. Its slow growth rate, late sexual maturity, and small litter sizes (7–19 pups) make it highly **vulnerable** to overfishing and slow to recover from population declines. Human activity, including coastal development, dredging, and mangrove destruction, poses an additional significant threat by destroying its **vital** shallow-water **habitats**.
- **Taxonomic confusion:** Historical confusion with other similar-looking species has made it difficult to collect accurate population data, complicating conservation and management efforts.

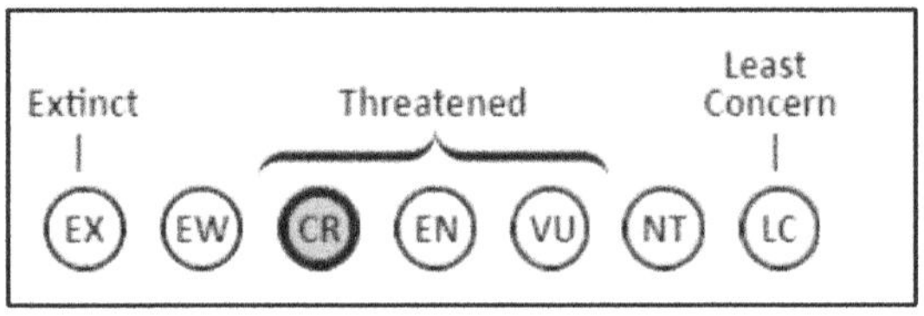

[1]

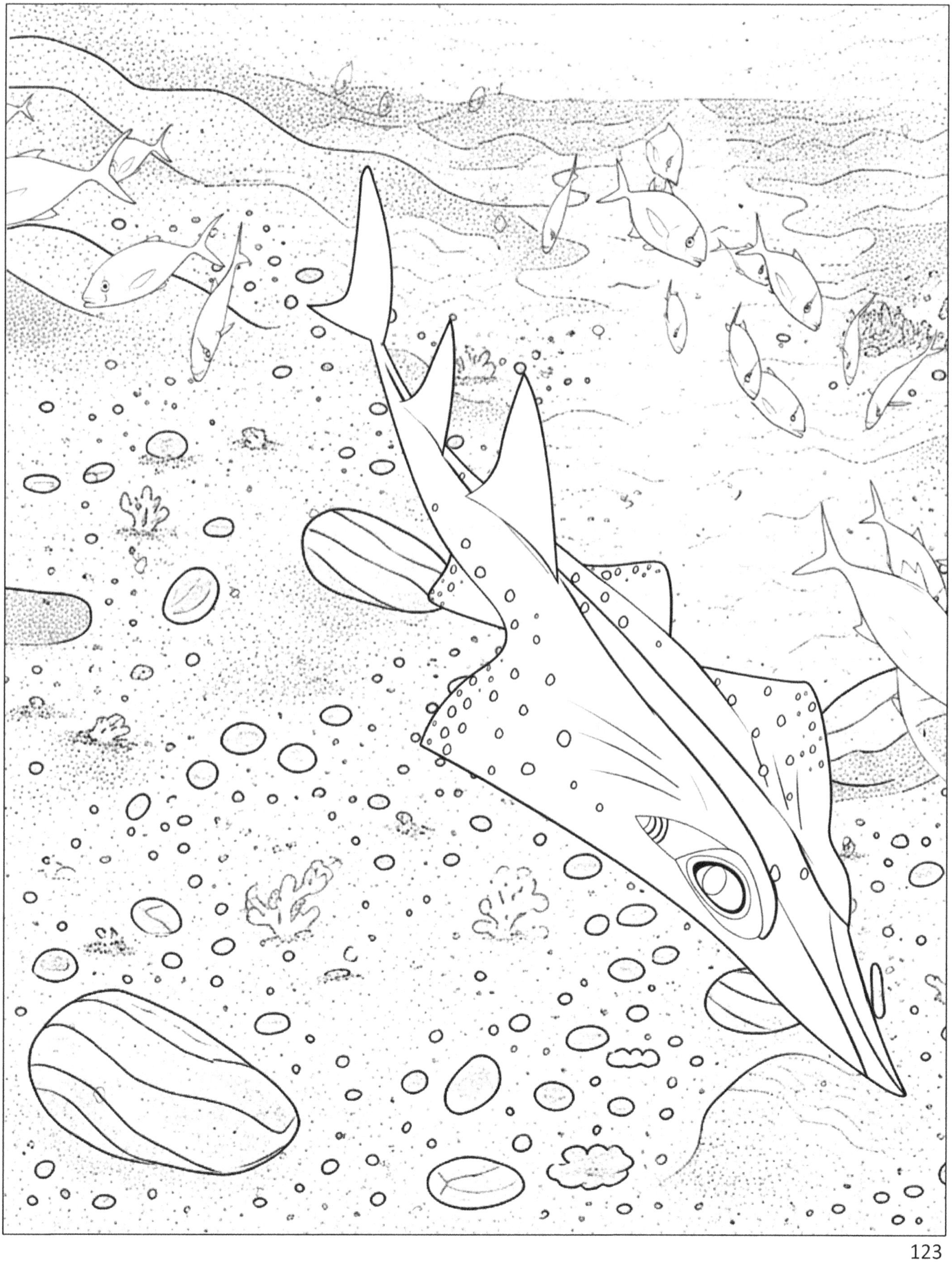

Bush Stone-curlew, Weerlo (Noongar)

Burhinus grallarius

- **Gangly** yet impossibly graceful, the bush stone-curlew or bush thick-knee is a **nocturnal**, ground-dwelling bird that makes its home in Australia's open forests, grasslands, mangroves, and salt marshes.
- Once widespread, the bush stone-curlew is now **rare** in most **regions** of the country. It is listed as endangered in New South Wales, and experts worry that in 10 or 20 years, it will be too late to prevent the **species** from becoming **extinct**.
- These birds are **terrestrial foragers**. By night, they feed on a selection of **prey** including insects, crabs, lizards, and small mammals, and by day they hide among the tall grasses and shrubs, folding their slender legs up under themselves as they rest.
- Their grey-brown colouration is distinguished by dark streaks, eyes are large and legs are long. They are capable of flight, but they rely on the **camouflage** of their **plumage** to **evade** detection during the day.
- Nicknamed the 'screaming woman bird', their high-pitched, drawn-out shrieks can be heard across the night as they try to contact each other. Often heard at night, the bird's **distinctive** call has been described as **akin to** the call of a screaming woman or baby. In many Australian aboriginal cultures bush stone-curlews have close associations with death, and features in many aboriginal stories across Australia.
- Bush stone-curlews breed throughout spring and summer, putting on an **elaborate courtship** dance involving standing with their wings outstretched, stamping their feet up and down, and keeping their tail upright.
- Once a bush stone-curlew finds a mate, this bond remains throughout its lifetime, which can last up to 30 years.
- During the **breeding** season, the bush stone-curlew will become particularly **territorial**, even with its own kind, and will try to ward off its competition with that powerful cry. It will also puff up its chest and spread its wings in an **aggressive** display to appear larger and more **formidable**.
- Its response to a **predato**r, however, is almost the exact opposite. If it catches wind of a fox, dingo, or goanna nearby, the bush stone-curlew will freeze, dead still, often committing itself to the strangest and most awkward of poses.
- Threats to the bush stone-curlew include **predation** by foxes and cats, **trampling** of eggs by cattle, clearance of woodland habitat for **agricultural** and **residential** development, modification and destruction of ground habitat through removal of litter and fallen timber, introduction of **exotic** pasture grasses, grazing and frequent fires, disturbance in the **vicinity** of nest sites.

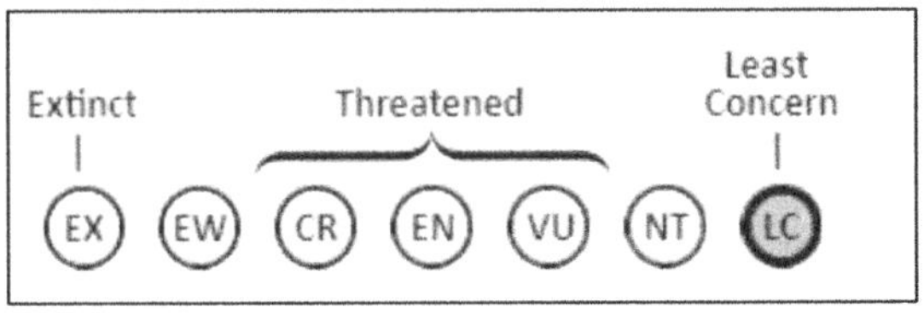

[1]

Star-leaved Grevillea

Grevillea asteriscosa

The star-leaf grevillea, is a **critically endangered** shrub from southwestern Western Australia known for its star-shaped leaves and bright red flowers that bloom from May to November. A key fact is its "Critically Endangered" status according to the IUCN Red List, though it is sometimes listed as "Priority Four" (rare, near threatened) by Western Australian authorities. Other interesting facts include its hairy, spreading branches, its preference for well-drained soils like gravel and **granite**, and that its specific name "*asteriscosa*" means "abounding in little stars," referring to its leaves.

- **Characteristics and habitat**
  It is a shrub with widely-spreading branches and **distinctive** star-shaped leaves that have three to nine sharply-pointed lobes. The flowers are bright red. It typically grows to a height of about 0.3–2.6m. It grows in heath and scrub in the Avon Wheatbelt and Mallee regions of southwestern Western Australia. It prefers well-drained soils in full sun. Flowers are arranged in groups of four to ten and appear between May and November.
- **Conservation and naming**
  It is listed as "Critically Endangered" by the IUCN and is sometimes referred to as "Priority Four" (rare, near threatened) by the Western Australian government. The species name, *asteriscosa*, is Latin for "abounding in little stars," a direct reference to its leaf shape. It was formally described by German botanist Ludwig Diels in 1904.

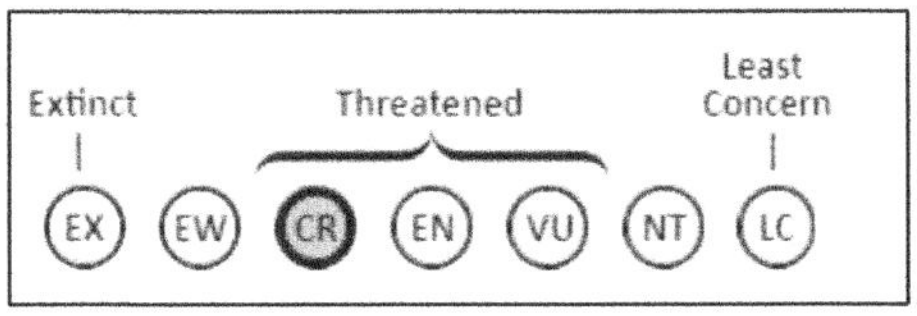

[1]

Children's Python

Antaresia childreni

The Children's python is known for being a docile, non-**venomous** snake that is popular as a pet due to its manageable size and easy care. It is a **nocturnal** hunter native to Australia that can be found in a variety of habitats, including woodlands and rocky areas. An interesting fact is that the female python coils around her eggs to incubate them by regulating their temperature.

- **Characteristics and behaviour**
  This python is known for its generally mild and docile temperament, making it a good choice for beginner snake owners. They typically grow to about 75cm, but can reach up to 1m.
  As a **nocturnal** animal, it hunts for prey such as small mammals, birds, and reptiles at night. Like other pythons, it kills its prey by **constriction** before swallowing it whole.
  The female python exhibits maternal care by coiling around her eggs to **incubate** them, which can take up to two months.
- **Habitat and diet**
  They are found in a variety of **habitats** across northern Australia, including **savannas**, woodlands, and rocky outcrops. They are **semi-arboreal**, meaning they spend time both on the ground and in trees or rock **crevices**. Adults primarily eat small **mammals** and birds, while **juveniles** might start with lizards.
- **Naming and conservation**
  The species was named after John George Children, a curator at the British Museum's zoological collection. They are listed as "Least Concern" and have a stable, possibly growing, population. Their small size and gentle nature make them a popular pet snake.

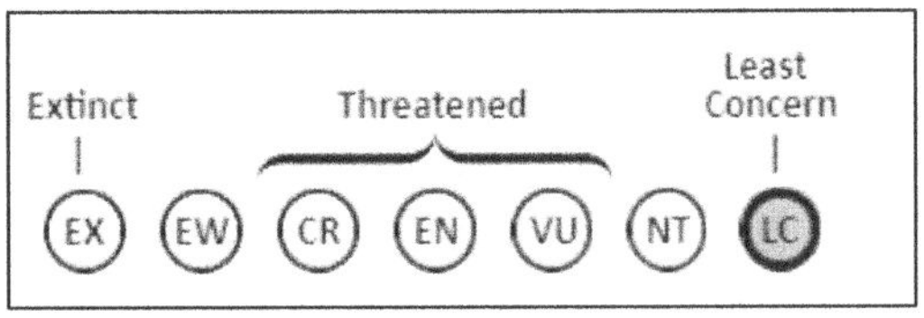

[1]

Lesser Long-eared Bat

Nyctophilus geoffroyi

The lesser long-eared bat is a widespread Australian microbat known for its impressive hearing, **unique** hunting techniques, and **adaptability** to **urban environments**.

- **Auditory adaptations and hunting strategy**
  The lesser long-eared bat has very large, ribbed ears (up to 25mm long) which are joined across its forehead. When at rest, the bat can fold these ears back. Instead of relying solely on loud **echolocation** like other bats, the lesser long-eared bat can use a very quiet "whispering" technique. This allows it to hunt for insects on the ground or on **foliage** without being heard by prey, particularly moths, which have developed ears to detect bat **echolocation**. This bat has a diverse hunting style, utilising **echolocation**, sight, and passive listening to pinpoint prey. It can grab insects in mid-air, snatch them from leaves, or pick them off the ground. It is one of Australia's slowest flying bats, moving at only about 4km/h while hunting. This low speed, combined with high **manoeuvrability**, is an **adaptation** for snatching prey from close range.
- **Physical traits**
  A small, Y-shaped groove can be seen behind the nostrils. This unique nose-leaf structure assists with **echolocation**. The bat has light grey-brown fur on its back, which is distinctly bicoloured (darker at the base), and a paler, sometimes white, belly.
- **Behaviour and reproduction**
  The lesser long-eared bat has a remarkable reproductive strategy. Mating occurs in April, but the female stores the sperm until ovulation and fertilisation take place in late winter or early spring (August/September). After a **gestation** period of 72–93 days, females often give birth to twins. The young are born between late October and November and can fly by December. To conserve energy during cold winter months, the bat can enter a state of **torpor**, which drastically lowers its **metabolic** rate and body temperature.
- **Habitat and distribution**
  The species is highly **adaptable** and lives across most of Australia, from **arid** deserts to alpine woodlands, rainforests, and **urban** areas. It is absent only from the tropical northeast coast and Cape York Peninsula. Though it uses natural tree hollows, fissures, and caves, the lesser long-eared bat is very comfortable **roosting** in man-made structures. **Roost** sites can include hollow walls, ceilings, unused roller doors, canvas awnings, and even old clothing. It is a common species in towns and suburbs, where it has successfully **adapted** to live alongside humans. This proximity, however, also makes it **vulnerable** to domestic cats.
- **Threats**
  Despite its **adaptability**, the bat is still **vulnerable** to threats like **habitat** loss and **degradation** from land clearing and forestry. The loss of large, old eucalypts is especially damaging, as these are preferred sites for **maternity colonies**. Due to its low-flying hunting style and **urban habitat**, it is frequently killed by domestic and **feral** cats. The use of **pesticides** is also a serious threat, as it can harm the bat's insect food supply.

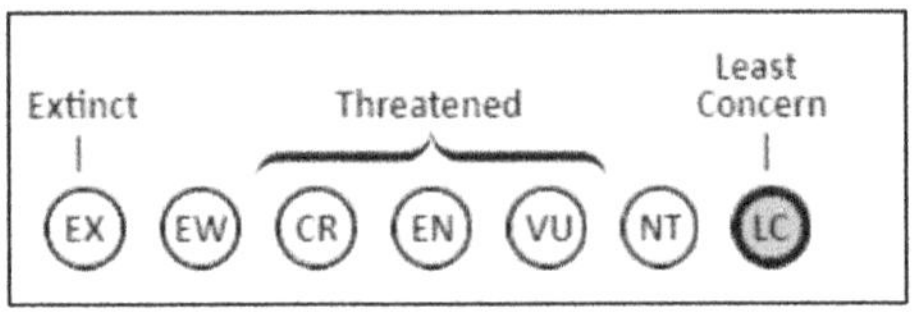

[1]

Microbats

General information about microbats

- Bats are the world's second biggest group of **mammals**, and scientists are still finding new species of little microbats today. Approximately 70% of bats are microbats.
- Microbats are small, and the average body is about the size of a mouse. They are often called 'flying mice'.
- Microbats weigh from 2 grams to 200 grams - a teaspoon of sugar weighs about 5-6 grams!
- All bats are **mammals**. **Mammals** are animals that care for their young after birth by feeding them milk. Most **mammals** give birth to live young. They all have fur or hair on their bodies. Bats are the only **mammal** that can fly.
- Bats' wings are long hands – like us, they have one thumb and four fingers.
- Bats are **nocturnal**. They wake up at dusk and fly out to feed and find water. They return to the **roost** early in the morning. Bats sleep or rest by hanging upside down, using a special claw on their foot to ensure they won't let go when they're asleep. By hanging upside down, they don't waste any energy.
- Microbats can reach speeds of up to 160km per hour!
- A microbat's diet largely consists of insects but can include small **mammals**, frogs, and fish. Microbats can eat up to half their body weight in insects each night, equivalent to hundreds of insects per hour.
- Hearing is the most developed of all the senses - they can hear sounds up to 200,000Hz (we can only hear frequencies up to 20,000Hz)!
- Microbats do not rely on sight to find food. Instead, they use a form of radar called **echolocation**. Microbats **echolocate** by bouncing sound waves off objects and listening to the echo. They also have a magnetic sense – they can perceive the Earth's magnetic field and use it for orientation when flying.
- Most Australian microbats **roost** in tree hollows or under bark, or they can be found in caves and cave-like structures. When there aren't enough 'natural' **roosts**, they can also find shelter in roofs or sheds.
- Microbats also have nurseries where the young are left in large groups while the mothers feed at night. In some areas, these nurseries can consist of millions of bats.
- Food can be scarce in winter, so microbats survive by either **migrating** to warmer regions or they **hibernate**. **Hibernation** is a state of **torpor** (inactivity). Microbats can slow their heart rate and lower their body temperature, sometimes to as low as 2°C.
- The oldest known bat fossil is 52 million years old. Research shows it could fly but not use **echolocation**. Bats have been used as food sources, are a feature in Dreaming stories, are found in cave paintings and modern art, and are **totem animals** for some **indigenous** groups.

- Loss of **habitat** is the single, largest threat to all bats. **Predators** include cats, birds of prey, snakes, and, sometimes it is thought, even other bats such as the broad-nosed bat (see page 132).
- Like other wild animals, some bats may carry **viruses** that can be **transmitted** to humans. The two major **viruses** that have been identified in some bats are the Australian Bat Lyssavirus and the Hendra Virus - both can be **fatal**!
- <u>If you find a sick or injured bat, **do not touch it**. Contact a local vet for details or your nearest wildlife carer</u>.
- Microbat **habitats** include:
    - Shelter – such as trees or caves
    - Cave-dwelling microbats don't always need huge caverns to **roost** in. Even small rocky outcrops and **crevices** will do. However, **maternity colonies** need more room and can be found in larger caves.
    - Food – **abundant** insects to eat.
    - Water – microbats will wet their belly and lick their fur to get extra water that they don't get from their food
- Man-made habitats:
    - Houses
    - Farmland
    - Buildings in cities
    - Abandoned mines
    - Holes in old concrete drains can provide **habitat** for microbats.
    - Cave-dwelling microbats will use tunnels and abandoned mine shafts to **roost** in.
    - Microbats can often be found in houses or other buildings where they are safe and warm, and no other natural **roosts** are available.
- How can we help?
  - Microbats need:
    - Forests, old trees, hollows.
    - **Abundant** food.
    - Places they won't be disturbed.
  - We can help by:
    - Stop clearing forests and old trees.
    - Installing bat boxes where old trees have been removed.
    - Reduce the use of **pesticides** around the home and on farms.
    - Know where microbats are **roosting** and leave them alone.
    - Learning more about microbats.
    - Telling others about microbats.

Western Bush Wallaby, Western Brush Wallaby, Kwoora (Noongar)

Notamacropus irma

The western brush wallaby also known as the black-gloved wallaby, is a speedy, **agile marsupial** native to southwestern Australia. They are **unique** among their kangaroo relatives for their more **diurnal** (day-active) habits and their recovery from near **extinction** due to successful fox control programs.

- **Unique characteristics**
  The western brush wallaby is easily identified by its striking white facial stripe running from its ear to the corner of its mouth. It also has black "gloves" on its hands, black feet, and black and white ears. These wallabies are exceptionally fast and **agile**, capable of weaving and sidestepping to **evade predators**. They run with their head held low and their long, crested tail extended **horizontally** for balance, an **adaptation** for open ground life. While most **macropods** are **nocturnal**, western brush wallabies are most active during the day, particularly in the early morning and late afternoon (making them **crepuscular**). They rest during the hottest parts of the day. The diet of this wallaby is not fully understood, but it consists of a variety of grasses, leaves, and shrubs. They can seemingly survive for long periods without needing to drink water. Like other **macropods**, they use a **unique** "**pentapedal**" **locomotion** at slow speeds. This involves placing their weight on their small forelimbs and thick tail while their powerful hind limbs swing forward.
- **Recovery from near extinction**
  In the early 1970s, the western brush wallaby population underwent a devastating decline of about 80%, primarily due to an increase in introduced red foxes. Researchers in the 1990s developed a successful baiting program using an environmentally friendly **toxin** to control the fox population. This effort proved so successful that the wallaby population recovered significantly. Following this recovery, the western brush wallaby's conservation status was officially downgraded by the IUCN from "Near Threatened" to "Least Concern".
- **Elusive and understudied**
  Despite their recovery, western brush wallabies remain one of Australia's less understood **macropods**. Many aspects of their behaviour, including their precise diet and **social** structure, have not been extensively studied. While often solitary, they can sometimes be seen in pairs or small groups when **foraging**. However, interactions with other wallabies are rare. Details of their **reproductive** behaviour are also not fully known. Females typically give birth to a single **offspring** in April or May, which spends up to seven months developing in the pouch. Conservation groups like Bush Heritage Australia are actively studying western brush wallabies using camera traps and GPS collars to learn more about their movements and home range.

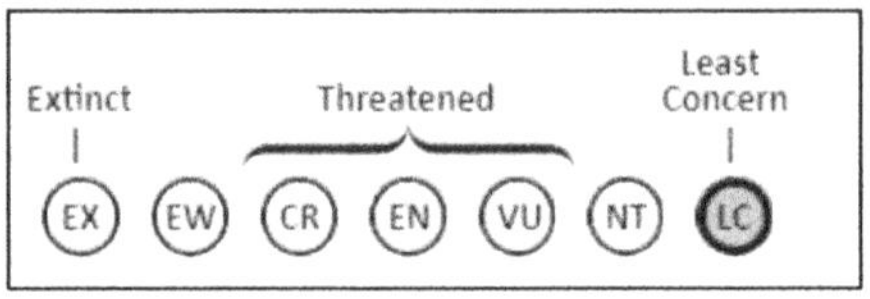

[1]

Quokka

Setonix brachyurus

Quokkas are known as the "happiest animals" due to their "smiling" appearance, but they are also able to climb trees, can survive for long periods without water by getting moisture from their food, and are related to kangaroos and wallabies. They are **nocturnal** and most active at dawn and dusk, are generally peaceful, but can be dangerous if threatened and should not be touched. Quokkas are vulnerable to extinction and have a **unique** reproductive ability where a new embryo will only develop if the previous **joey** has died.

- **Physical characteristics and abilities**
  Their smile-like expression is an accident of evolution, caused by the shape of their mouth. They are related to kangaroos and wallabies and are **marsupials**. Quokkas are able to climb trees up to 2m high to find food. They are primarily **nocturnal** but can be seen during the day, especially in the early mornings and late afternoons.
- **Diet and survival**
  Their diet consists of grasses, leaves, and succulents. They can survive in dry environments by getting all the moisture they need from the plants they eat. They can survive for extended periods without food or water.
- **Reproduction**
  Like other marsupials, their babies, called **joeys**, are born tiny and develop in their mother's **pouch** for several months. They have a unique reproductive strategy where a new **embryo** will only develop if the previous **joey** dies, ensuring the mother is almost guaranteed to have at least one healthy baby per year. The quokka is a protected species and is considered vulnerable due to its population size and threats like habitat loss and predation.
- **Behaviour and interaction**
  Quokkas are **social** animals and often live in small family groups. While they may seem friendly and "cuddly," they are wild animals with sharp teeth and claws. It is important not to touch or feed them, as this can make them ill or cause mothers to abandon their young.
- **Medical research:**
  Their unique characteristics have made them valuable in medical research, particularly for studying muscular dystrophy.

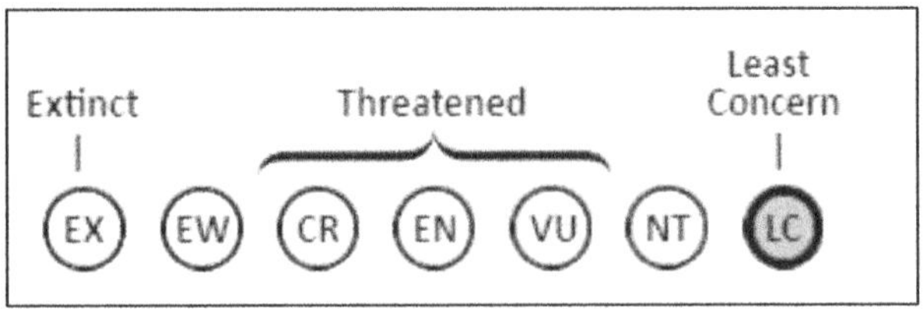

[1]

Diagram of banksia inflorescence

From Australian Botanic Gardens – Artist: Murray Fagg (available as .pdf download)

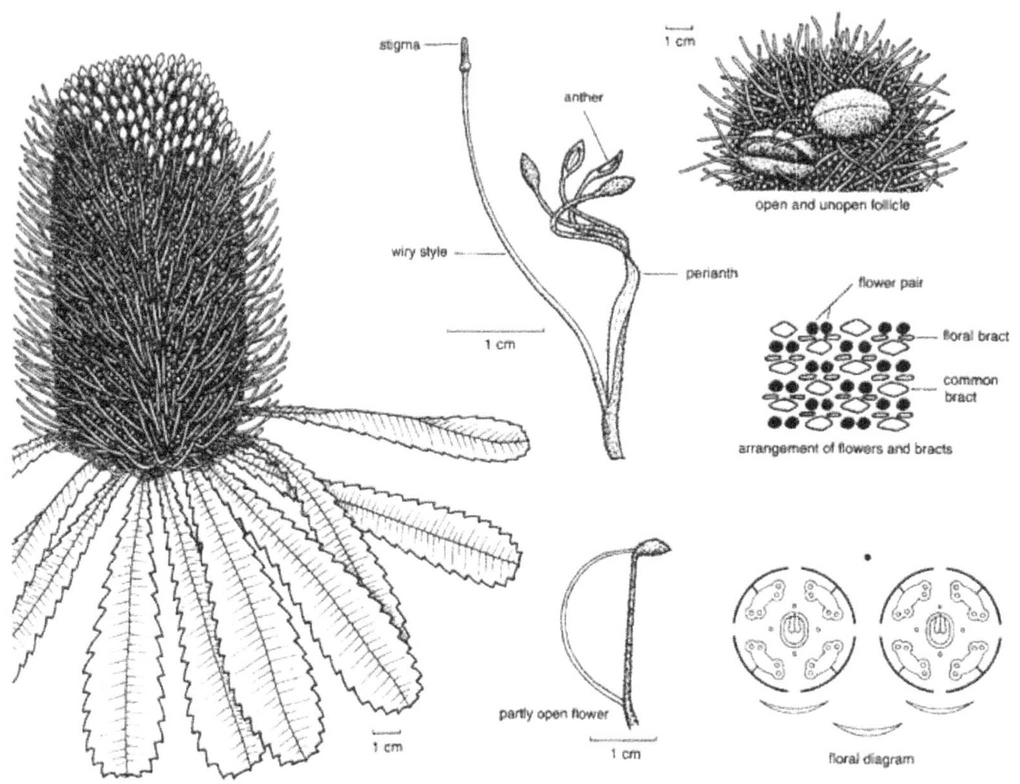

Parts of a flower:

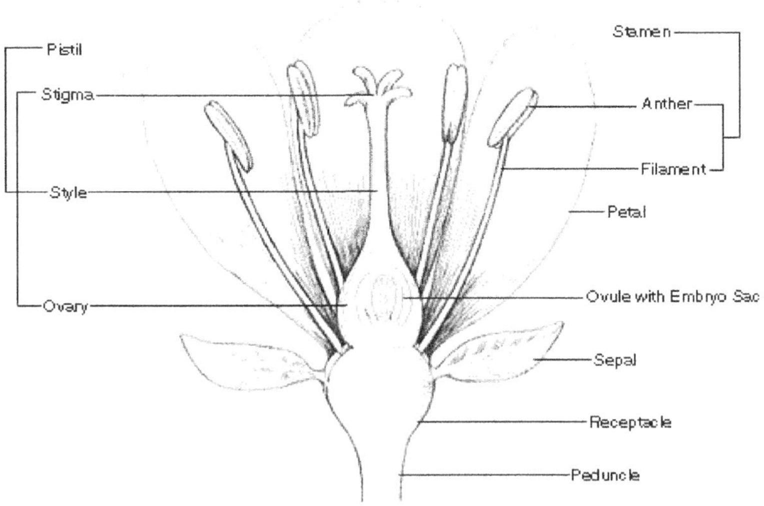

Peduncle: The stalk of a flower.

Receptacle: The part of a flower stalk where the parts of the flower are attached.

Sepal: The outer parts of the flower (often green and leaf-like) that enclose a developing bud.

Petal: The parts of a flower that are often conspicuously coloured.

Stamen: The pollen producing part of a flower, usually with a slender filament supporting the anther.

Anther: The part of the stamen where pollen is produced.

Pistil: The ovule producing part of a flower. The ovary often supports a long style, topped by a stigma. The mature ovary is a fruit, and the mature ovule is a seed.

Stigma: The part of the pistil where pollen germinates.

Ovary: The enlarged basal portion of the pistil where ovules are produced.

Just for fun: fins and feathers!

Bony fish anatomy:[37]

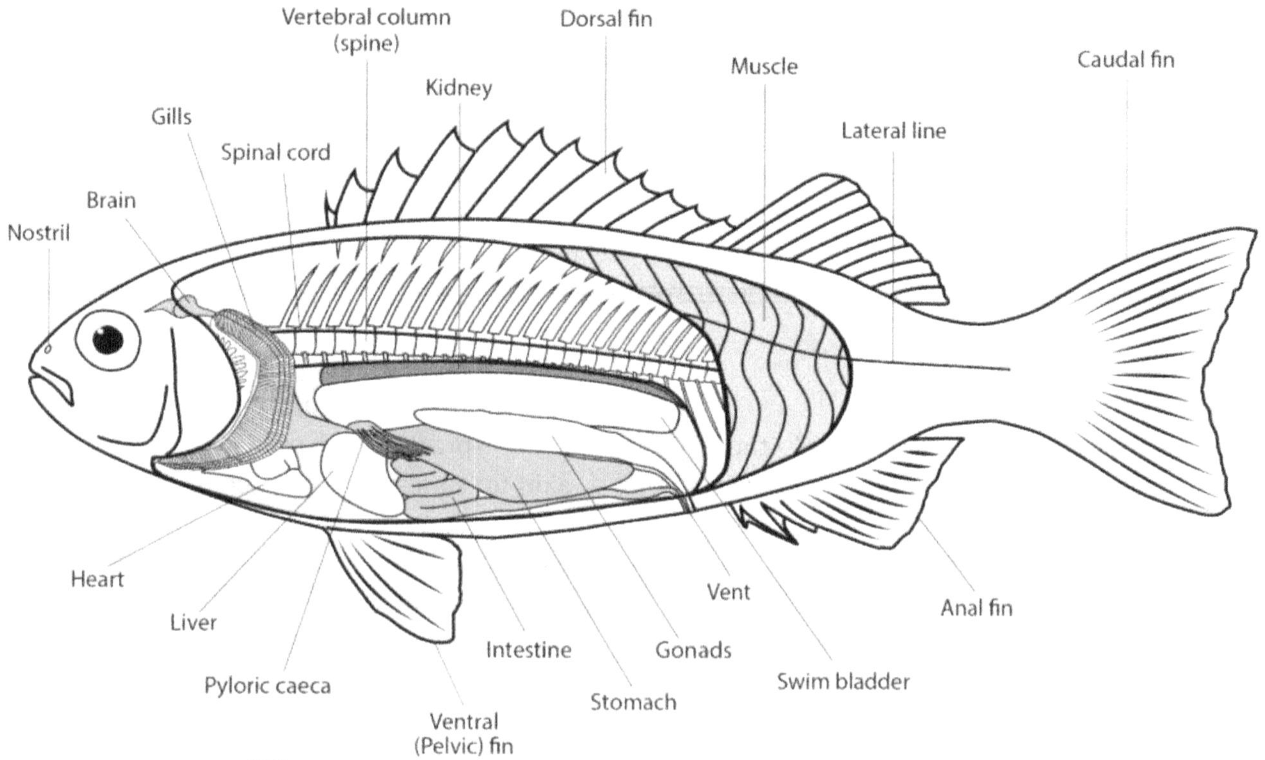

Basic bird anatomy:[38]

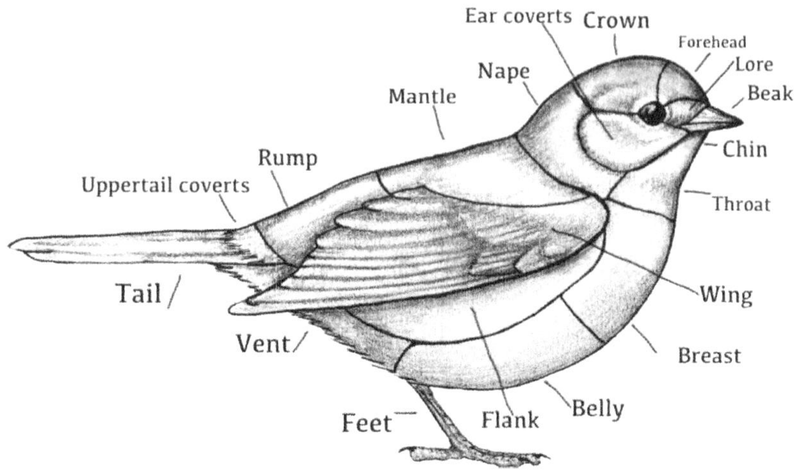

GLOSSARY – All those big or unusual words! …

Some of these words are really unfamiliar, even to me as the author of this book, but I have included them because I think they could be useful for any person, child or adult, who has an interest in nature. For the most part, a synonym (word that means the same thing) has been used to try to keep these explanations as simple as possible. Definitions and synonyms from Dictionary.com and Thesaurus.com have been used on occasion. These definitions are not intended to be 100% perfect, but to give younger readers (and perhaps the adults reading with them) some other ideas to help them understand and explain what may be unfamiliar words, in the context of this book! Not all terms will be found in the text; however, they are still good terms to learn for any enthusiastic nature lover.

---

abdomen: belly, gut, midsection
aboriginal: original, earliest, native
aboriginal mythology: The "Dreaming" - the rich oral traditions, stories, and beliefs of indigenous Australians, passed down through generations
absorb: take in, soak up
abundant: ample, plenty of, extensive
abyssal (ocean): relating to or denoting the depths or bed of the ocean, especially between about 3000 and 6000m down
abyssopelagic (ocean zone): abyssal zone
accumulate: to gather or collect something over time, causing it to gradually increase in quantity
achene (botany): a small, dry one-seeded fruit that does not open to release the seed
acid, acidic: substance like lemon, sour
acoustic: relates to sound, hearing, or the science of sound
acrobat, acrobatic/s: like gymnastics, athletic
adapt, adapted: change to suit, alter
adaptation/s: adjustment/s, alteration
adept: highly skilled or proficient at something
adhere, adhesive: stick, sticky like glue
adjoin/s, adjoining: be next to and joined with
aestivation/aestivating: a state of inactivity and lowered metabolic rate that animals enter during hot, dry periods to survive
aggressive: ready or likely to attack or confront
agile/agility: athletic, quick, energetic
agriculture, agricultural: the science or practice of farming
akin to: similar nature or character; related by blood
albino, albinism: pale or white colouring, usually with pink eyes
alga, algae: plants living in water
altitude, altitudes: height in relation to sea or ground level
ambush: trick, trap, a surprise attack
ambush hunter, ambush predator: an animal that prefers to capture or trap prey by stealth or strategy
amphibian: an animal that can live both on land and in water, cold-blooded vertebrate animal
amphidromous: organisms, typically fish, that regularly migrate between freshwater and saltwater environments, but not for breeding purposes
anal: related to or near the anus
analysis: study, test, investigate
anatomy/anatomical: structure of a plant or animal, the way it is built
angler: an animal/person who catches fish
annual, annually: yearly, once a year
antenna, antennae: feelers
antennules: small antenna, especially one of the foremost pair of a crustacean
anterior (anatomy): towards the front
anther: the part of the stamen of a flower that produces and contains pollen and is usually borne on a stalk

antivenom: a substance used to neutralise or counteract the effects of a venom
anus: the opening at the end of the digestive system
appendage/s: a part that is attached to a larger body
apex predator: a meat eating animal that preys on other animals but has no natural predators of its own, at the top of the food chain
aquarium, aquaria: fishbowl, fish tank
aquatic: occurring in water
aquatic vegetation: plants that rely on wet environments
arboreal: living in or among trees
archaea: microorganisms which are similar to **bacteria** in size and simplicity of structure but radically different in molecular organisation. They are now believed to constitute an ancient group which is intermediate between the **bacteria** and **eukaryotes**
arid (of land or a climate): having little or no rain; too dry or barren to support vegetation
arthropod: an animal with a segmented body, jointed limbs and usually a shell including insects, spiders, crustaceans
arsenic: a highly toxic metalloid that is naturally found in the earth's crust
artificial: not natural, made by humans
asexual/asaexually: independent of sexual processes
assimilate: ingest/digest, soak up
Austral spring/summer: a season in the Southern hemisphere
awn: a stiff bristle, especially one of those growing from the ear or flower of barley, rye, and many grasses
axis: an imaginary line that goes through the middle of something
bacteria/bacterium: a member of a large group of unicellular microorganisms which have cell walls but lack organelles and an organised nucleus, including some that can cause disease.
bacterivore/bacterivorous: an organism which obtains energy and nutrients primarily or entirely from the consumption of bacteria.
barb: a sharp projection near the end of an arrow, fish hook, or similar object, which is angled away from the main point so as to make extraction difficult; a fleshy filament at the mouth of some fish such as barbel and catfish
barbed: having a barb or barbs
barbel: a fleshy filament growing at the mouth of some fish
bask: to lie or relax in pleasant warmth
bathypelagic (ocean zone): midnight zone
benthic: relating to or occurring on the bottom of a body of water
bill (of birds): beak; the hard, pointy part of a bird's mouth
billabong/s: small lake usually formed when the path of a creek or river changes leaving a dead end
biocide/s: a substance that kills life
biodiversity: differences in biology within plant and animal species
bite force: force applied between the upper and lower mouth structures, jaw and/or teeth
bivalve: any mollusc with two shells hinged together
bog: a type of wetland with spongy, peat-rich soil that accumulates partially decayed plant matter
botany: the scientific study of the physiology, structure, genetics, ecology, distribution, classification, and economic importance of plants
bracken: type of big, bushy, green fern
brackish: water that is saltier than fresh water but not as salty as seawater
brood: a number of offspring, clutch, youngsters

brood chamber: a chamber, typically underground, in which young (animals or insects) are born and raised until old enough to venture outside
brood pouch: a pouch where eggs develop and hatch
brumination: a natural state of dormancy that reptiles enter during cold weather or other stressful conditions, such as prolonged heat or drought, to conserve energy
bunyip: the bunyip is a creature from the Aboriginal mythology of southeastern Australia, said to lurk swamps, billabongs, creeks, riverbeds, and waterholes
buoyant/buoyancy: the upward force exerted by a fluid (like water or air) that opposes the weight of an object immersed in it, making objects float or rise
burrow: hole or tunnel dug by an animal, usually for a home or hiding place
buttress roots: large, wide roots sometimes known as plan roots on all sides of a tree
calligraphy: decorative handwriting or handwritten lettering
camouflage: disguise, mask, concealment
canine/s: a pointed tooth between the incisors and premolars of a mammal, often greatly enlarged in carnivores
canopy: the upper part of a tree or plant group that is made up of leaves and branches. It is also known as the overstory
capillary action: the ability of a liquid to flow in narrow spaces, like a thin tube, without being pulled by external forces like gravity
captive breeding: bred in captivity
captivity: kept in a place like a zoo or a cage
carapace: a bony or chitinous shell, test or shield covering some or all of the dorsal part of an animal, e.g. turtle
carnivore/carnivorous: meat eating
carotenoid: nature's food colouring! They're the bright yellow, orange, and red pigments that give fruits and vegetables their colourful appearance
carpal: relating to the bones forming the wrist
carrion: dead and rotting flesh of an animal
cartilage: strong, bendy, and rubbery tissue
cartilaginous tissue: made out of cartilage like our ears and nose tips
caudal (anatomy): of, relating to, or being a tail; directed toward or situated in or near the tail or posterior part of the body
cauline (botany): relating to or growing from a plant stem, particularly the upper part of the stem
cavity: an unfilled space within a mass (especially a hollowed-out space)
ceremony: a formal event with a specific purpose
characteristic: distinctive, singular feature
chemical cues: hints that are chemical, could be particular smell
chenopod grassland: *Chenopodioideae* are annual or perennial herbs, subshrubs, shrub or small trees
chromatophore: tiny colour-changing cells under an animal's skin
chrysalis: the hard-shelled pupa of a moth or butterfly
chytrid fungus: an infectious fungal disease which can be fatal to amphibians
chytridiomycosis: is an infectious disease of amphibians – see **chytrid fungus**
circulate: move in a circle, pass from place to place, spread
circumference: the distance/measurement all the way around an object/circle
cite, cited: quote, name, reference
classify, classification (scientific): the sorting of things into groups based on how they are alike
clime/s: the average weather conditions of a particular area

cloaca/cloacal: a shared exit for bodily waste and reproductive function
clutch (of eggs):total number of eggs/young in one nesting
coastal dredging: excavating material from a coastal water environment
co-exist: live or exist together
colony/colonies/colonisation: group, settlement, swarm
colour morph: a specific body colouration variation
combatant/s: fighter, battler
commerce: business, trade
commercial: merchant, business, of trade, public
commercial port: a place for ships to load or unload things for buying and selling
communal: shared or used by all members of a group
compact: something is closely packed or a small size, taking up little space
complex: complicated, convoluted
condense: to make something shorter, more compact, or more concentrated
cones (eyes): specialised cells that allow us to see colour and fine details, particularly in bright light
configuration: composition, shape, structure
congregate: gather into a crowd or mass
conservation: saving and protecting
conspicuous: easily seen or noticed; standing out from a group
constrict/constriction: to become tighter or narrower, or to make something become tighter or narrower
construct: build or put together
consume: eat, drink or use something up
contamination: infection, pollution, corruption
continent: big, solid area of land – Earth has seven continents
co-operative: working with others to help make things happen
copepod: a tiny, shrimp-like crustacean with "oar feet" that lives in nearly all aquatic environments
copious: abundant, plentiful, or large in quantity or number
coronal (anatomy): dividing body into front and back
counterpart: equals, fellow, like, twin
courting, courtship: set of display behaviours in which an animal, usually a male, attempts to attract a mate
cranial (anatomy): towards the head
creature: a living animal or living insect
creche: a group of young cared for communally
credible: believable, reasonable, likely
creek: small stream or river
crepuscular: resembling or relating to twilight
crevice: slit, crack, cleft, gap
critical: very important
critically endangered: facing an extremely high risk of extinction in the wild
critter: informal word for any living creature
cross-pollination: the transfer of pollen from the flower of one plant to the flower of a different plant of the same species
crucial: extremely important or necessary
crustacean/s: mainly aquatic arthropod typically with body covered with a hard shell or crust, such as lobsters, shrimps, crabs, barnacles, wood lice
cryptic: secret, mysterious, hidden
ctenoid scales: scales with spines or comb-like teeth along their free edge
cull/culling: reduce the population of (a wild animal) by selective slaughter
culprits: wrongdoer, offenders
cultivate: breed, raise, tend to, plant
cultural significance: of some value for past, present or future generations
culvert/s: tunnel carrying a stream or open drain under a road or railway
currency: money, legal tender
cyanobacteria: formerly known as blue-green algae, are photosynthetic microscopic organisms that are technically bacteria

cygnet: young swan
cylinder, cylindrical: 3D shape like a tube with circles on either end
debris: scattered pieces of rubbish or remains
decade: 10 years
decline/declining: drop, gall, decay, weakening
decorticate/decorticating: remove the bark, rind, or husk from.
decurved: curved downward or bent downward
deep (anatomy): further from the surface
defence mechanism/s: adaptations that help avoid predators or mitigate adverse effects.
defense, defensive: protecting or defending against something
deforestation: the action of clearing a wide area of trees
degrade, degradation: reduce in quality, lessen, weakening, deterioration
demersal: living or found near the bottom of a body of water
den: the lair of a wild animal
dense: thick, heavy, impenetrable, solid
derivation (words): where the word/s came from
destruction: breaking something apart, causing so much damage it can't be used or fixed, or even removed
detritus: small pieces of disintegrated material such as organic matter
diagnosis/diagnostic: the process of identifying a disease, condition, or injury by looking at its signs and symptoms
difference between endemic and native? An endemic species is a native species but found only in a particular area, large or small
digest: to break down food into smaller pieces to use for energy and growth
digestive system: the parts of the body that work together to turn food and liquids into the building blocks and fuel that the body needs for energy and growth

digit/s: finger, thumb or toe
dimorphic/dimorphism: two different forms
dioecious: having the male and female reproductive organs in separate individuals
disease: illness or sickness characterised by specific signs or symptoms
disperse/dispersal: to spread out or move in different directions
dissipate: disappear, waste or fritter away
distal (anatomy): further from the centre
distinction/s: noticing the difference between two things
distinguish, distinguishes: mark off as different
distinctive, distinctively: having a special quality, style, colour, etcetera
distribute, distribution: divide into, disperse through a space or over an area, spread, scatter
diurnal: of or during the day.
diverse/diversity: different or having variety
dominant/dominance: more strong than others
dormant: inactive or not active, but capable of becoming active later
dorsal, dorsum (anatomy): of, relating to, or situated at the back, or dorsum of
drainage: removing excess water from an area
drey: the nest of a possum or certain other animals, like a squirrel
drought: a long and abnormally dry spell
drupe: a type of fleshy fruit that contains a single hard seed enclosed in a stony layer
duck blind: shelter, often camouflaged with reeds and grasses, for hidden observation
dusk: time of day when the sun is setting and it's starting to get dark, just before night
dusky (of colour): of a darker colour, shadowy
dwindle: become less or smaller gradually
echolocation/echolocate: the location of objects by reflected sound
ecosystem: a geographic area where plants, animals, and other organisms, as well as weather and landscapes, work together to form a bubble of life
ectomycorrhizal: a symbiotic association of fungi with the feeder roots of higher plants

edible: fit to be eaten as food, eatable
eel: long snake-like fish
elaborate: complicated, detailed, **complex**
elliptic (botany): the shape of structures like leaves, being widest at or near the middle and tapering to both ends, similar to an oval
elusive: tricky, difficult to catch, shy
endemic: native species found only in a particular area
embryo: baby in its very early stages, before it's born; animal or plant that's just starting to grow and develop
embryonic diapause: a temporary suspension of embryonic development at the blastocyst stage, acting as a reproductive strategy to delay birth until conditions are favorable for offspring survival
endangered: there is a very high risk of extinction due to rapid population declines of 50 to more than 70% over the previous 10 years, a current population size of fewer than 250 individuals
environment, environmental: the natural surroundings that make up the world we live in
EPBC: Environment Protection and Biodiversity Conservation Act 1999 (EPBC Act) is Australia's main environmental law
ephemeral: lasting for a very short time; fleeting or transitory
epic: very large, grand, or impressive
epicormic buds: dormant buds that lie beneath the bark of a tree
epipelagic (ocean zone): sunlight zone
estimate, estimated, estimating: guess, projection
estuary/estuarine: an area where a freshwater river or stream meets the ocean
eukaryotes: any cell or organism that possesses a clearly defined **nucleus**
European settlement: 1788-1810 early European settlement
evade: avoid or get away from something, often by being clever or sneaky
evolve, evolved: grow, mature, advance
evolution, evolutionary: change, expand, progress, transform
excavate: dig out, burrow, uncover
excretion/s: discharge by excreting
exotic: of foreign origin or character, introduced from abroad
extensive, extensively: covering a large area
extinct/extinction: having died out, no longer existing
extreme weather events: occurrences of unusually severe weather or climate conditions that can cause devastating impacts on communities and agricultural and natural ecosystems
exudate: substance coming out, ooze
fatal/fatalities: causing death
fauna: all the animal life in a specific region or time period
faunal region: any of six or seven areas of the world defined by animal geographers on the basis of their distinctive animal life
feral: wild, having reverted to a wild state
fertilised, fertilisation: make productive, capable of development
filament/filamentous: a single, thin, thread-like object or structure
fire regimen: pattern, frequency, and intensity of the bushfires and wildfires that prevail in an area over long periods of time
flank/s: side between ribs and hips
flock: group of animals, especially birds like geese or chickens, or sheep and goats, that stay together
flora: all the plant life in a particular region, time, or environment

foliage: the leaves of a plant or tree collectively, or leaves in general
folivore: an animal or organism that primarily eats leaves
follicle/follicles (botany): dry fruit formed from a single carpel that splits open along one side to release its seeds
forage, foraging: look for food
formidable: something that is really strong, impressive, or even a little scary, but also something you might respect
fossil: the preserved remains or traces of a living thing that lived long ago
fragment, fragmented, fragmentation: broken off, scattered, isolated, separated
frugivore/frugivorous: an animal that thrives mostly on raw fruits or succulent fruit-like produce of plants such as roots, shoots, nuts and seeds
fungus, fungi, fungal: simple organism, or living thing, that is neither a plant nor an animal
gait: the way a person or animal walks or moves, including the pattern, style, and rhythm of locomotion
gangly: tall and thin and moving with a loose-jointed awkwardness
gemma/gemmae (botany): a single cell, or a mass of cells, or a modified bud of tissue, that detaches from the parent and develops into a new individual
generalist feeder: feeds on a wide variety of things and thrive in various environments
generic: referring to all members of a genus
gene/s: the basic unit of heredity. This means that genes determine what traits are passed down from a female and male to their offspring. Genes exist inside the cells that make up living things
genetic/s: traits carried in genes, which are like special instructions for making a body
genus: the major division of a family, usually of more than one species
geology/geological: the scientific study of Earth, including its physical features, its history, and the processes that have shaped it, such as rock formation and plate tectonics
germinate/germination: when a seed starts to grow into a plant
gestation: the process or period of developing inside the womb between conception and birth
gill: an organ used by aquatic animals breathe by extracting oxygen from water and releasing carbon dioxide
gizzard/s: a muscular organ in the digestive tract of some animals, like birds, that grinds up food
glabrous: smooth and hairless
gland/s: an organ in the human or animal body which secretes particular chemical substances for use in the body or for discharge into the surroundings
global/globally: worldwide, comprehensive
globose: having the shape of a globe or sphere; round
glossy: shiny, lustrous, polished
graffiti: writing or drawings, scribbled,scratched, or sprayed, usually illicitly
granite: a hard, light-colored igneous rock that forms from the slow cooling of magma deep underground
granivore/granivorous: an animal that primarily feeds on seeds and grains
granular: something that is made of tiny grains or particles, like sand or sugar, or that looks like or feels like it is made of those small bits
graze, grazing:eating, feeding, nibbling, munching
groundwater: water below surface of ground
habitat/s: place where usually found, natural environment of an organism
harem: a group of female animals sharing a single mate
harvest, harvested: gather food
hatch, hatching: baby animal coming out of its egg, or that an egg is breaking open to release the baby

hatchling: young bird, reptile or fish just hatched from egg
haven: a place of safety and refuge
heath/heathland: an area of open uncultivated land, typically on acid sandy soil, with characteristic vegetation of low-growing plants, shrubs and grasses
hectocotylus: a specialised arm on male cephalopods (like octopuses) that is adapted for reproduction
herbicide/s: a substance that is **toxic** to plants
herbivore/herbivorous: an animal that primarily eats plants
herper/s: a person who is interested in reptiles and amphibians, or herps
hibernate, hibernation: spend the winter in a dormant state
hinder: make harder or slower to do, make difficult
honeydew: sugary material secreted by some insects such as aphids, scale
horizontal: parallel to the horizon
host: an organism that provides shelter and nourishment to another organism, which can be a parasite, virus, or microbe
host plants, host trees: plants that an organism lives on and lives off (parasitic or semi-parasitic plants)
hostile: aggressive, warlike
human disturbance: people disturbing the environment: for example - tree removal. road building construction near shorelines and streambanks
humid, humidity: moisture or water vapour in air
hydrology: the branch of science concerned with the properties of the earth's water, and especially its movement in relation to land.
iconic: representative of
illegal: against the law, not legal
imminent: likely to happen very soon
imperilled: put in danger or at risk
immature: something that is not fully developed or grown
immune to: not being affected by something
impoundment (of streams): damming, enclosing
incisor/s: a narrow-edged tooth at the front of the mouth, adapted for cutting
incubate, incubated, incubation: brooding
indigenous: having originated in and being produced, growing, living, or occurring naturally in a particular region or environment
indigenous people: First Nations people
inferior (anatomy): below, lower than
infertile: unable to reproduce naturally
infiltrate/infiltration: the process of cells or substances moving across a barrier, typically a tissue barrier, into a place they are not normally found, or in which they are typically found in lower concentrations
inflorescences: the complete flower head of a plant including stems, stalks, bracts, and flowers
influx: the arrival or entry of many things
inhabit, inhabits: live in, occupy, populate, reside
innate: a quality or ability that exists from birth rather than being learned
instar: a developmental stage between moults, or shedding of the exoskeleton
interlock: to connect or join two or more things so that they are mutually dependent
insecticide: an agent (usually chemical) that destroys insects
insectivore/insectivorous: a carnivorous animal or plant which eats insects
intertidal: between high and low water marks
intimidate/d: scare, frighten, bully, terrify
introduce/introduced (of species): plant or animal that wasn't originally there but was brought in by humans, either on purpose or by accident
invasive: threatening a plant or animal community, by spreading prolifically
invertebrate/s: animals without backbones

iridescent: something that changes colours when you look at it from different angles

iridophores: an iridescent chromatophore

iris (of eye): the coloured tissue at the front of the eye that contains the pupil in the centre

isolation, isolated, isolate: detach, separate, place apart

IUCN: International Union for the Conservation of Nature

joey/s: young marsupial animal

juvenile/s: youth, junior, infant, young

keel (biology): a perpendicular extension of a bird's breastbone, to which wing muscles anchor. Keel (slug), a morphological feature on the back of some slugs. Caudal keel, in fish. Keeled scales, reptile scales that have a ridge down the centre, rather than being smooth.

keystone species: an organism that has a disproportionately large impact on its ecosystem relative to its abundance

kinetoplast, kinetoplastids: a group of flagellated protozoans that are distinguished by the presence of a DNA-containing region, known as a "**kinetoplast**", in their single large mitochondrion

labellum: a central petal at the base of an orchid flower, typically larger than the other petals and of a different shape

lactate, lactating, lactation: producing and secreting milk from the mammary glands

lagoon: a shallow body of water, usually saltwater or brackish, separated from a larger body of water like the ocean by a natural barrier such as a sandbar, barrier island, or coral reef

lamina (biology/botany): a thin layer or plate

land reclamation: the process of creating new land from oceans, seas, and rivers, or improving land that is unsuitable for use to make it suitable for agriculture, construction, or other purposes. It can involve creating new land from water bodies or restoring degraded land, such as that damaged by mining or natural phenomena

lantana thicket: compact shrub which can grow in dense thickets, areas of trees and bushes growing close together

larva, larvae: immature, wingless, feeding stage of an insect that undergoes complete metamorphosis

lateral (anatomy): towards the side

laterite: a reddish, iron- and aluminum-rich soil and rock formed in hot, wet climates from the intense weathering of underlying rocks

legume (botany): a dry fruit (a pod) that typically splits open on two sides at maturity, and it is produced by plants belonging to the *Fabaceae* family (also known as *Leguminosae*), which includes well-known edible species like beans, peas, and lentils

lichen: a complex, plantlike organism formed by a symbiotic relationship between a fungus and a photosynthetic partner, which can be an alga or a cyanobacterium

life cycle: the series of changes in the life of an organism

lifespan: the period of time of existence from birth to death or the end of useful function

lignotuber: a woody swelling at the base of a plant that contains dormant buds, nutrients, and carbohydrates

ligule (botany): a narrow strap-shaped part of a plant, especially a membranous scale on the inner side of the leaf sheath at its junction with the blade in most grasses and sedges

litter: number of young born in one birth
littoral: relating to or situated on the shore of the sea or a lake
littoral rainforest: a closed forest community that grows near the ocean, usually within 2km
local, locally, localised: restricted to one place
locally extinct: no longer present in a specific geographic region but still exists elsewhere
locomotion: the act of moving from one place to another
longitudinal/longitudinally: running lengthwise rather than across.
lore: the region between the eyes and nostrils of birds, reptiles, and amphibians
macro- (prefix): large; large-scale.
macroalga/e: a large, multicellular form of algae that is visible to the naked eye, commonly known as seaweed
macropod: a marsupial with large hind feet, such as a kangaroo, wallaby, or quokka
mammal: warm-blooded vertebrate having hair or fur, producing milk for the young and giving birth to live young – exceptions are the **monotremes.**
manoeuvrable/manouevrability: easily moved, steered, or controlled into different position
mantle (biology): a soft, fleshy outer covering of molluscs and brachiopods that is the body wall of the animal
Māori people: indigenous Polynesian people of mainland New Zealand
marine: of or relating to the sea
marsh/marshes/marshland: type of wetland ecosystem characterized by poorly drained mineral soils and by plant life dominated by grasses
marsupial: animals in the order *Marsupialia* which have a pouch for their young
marsupium: pouch found in female marsupials
maternity: the period during pregnancy and shortly after childbirth
maternity colonies: group of pregnant or lactating female bats that gather together in a roost to give birth, nurse, and raise their young
matriarch, matriarchal: a social system in which positions of power and privilege are held by females
mature, maturity: fully developed, having completed natural growth and development
mechanism: the fundamental processes involved in or responsible for an action, reaction, or other natural phenomenon
medial (anatomy): towards the middle
melody/melodious: sounding pretty and musical
membrane/s: sheath, sheet, film of tissue
mesial: being or located in the middle or a median part
mesopelagic (ocean zone): twilight zone
meatbolic, metabolism: the chemical reactions in cells that change food into energy
metamorphosis: change in life cycle e.g. pupa to caterpillar to butterfly
methodical: orderly, structured, deliberate
micro- (prefix): very small, tiny
microalgae: microscopic algae invisible to the naked eye; phytoplankton typically found in freshwater and marine systems, living in both the water column and sediment. They are unicellular species which exist I ndividually, or in chains or groups
microendemic: species with very small geographic distributions (ranges)
microorganism: an organism that is so small it can only be viewed using a microscope. Also called microbes, these tiny living things are found everywhere in water, soil, air, and the human body, with examples including bacteria, fungi, and protozoa
microphyte: a microscopic plant
microscopic: incredibly small, unable to be seen without a microscope

migrant: drifter, traveller, wanderer
migrate, migration, migrating: going from one country, region or place to another
migratory: characterised by migration
mimic, mimicking, mimickry: copy, simulate, imitate
mitochondria, mitochondrium: the tiny powerhouses inside our cells! They break down food we eat to give us energy, so we can move, think, and do everything else we do...think of them as little energy factories
moist: slightly wet, damp
moisture:the presence of a liquid, especially water, often in small amounts
molar: the back teeth in the mouth, shaped like flat, grinding surfaces
mollusc/s: animal that has a soft body, no spine, and is often covered with a shell
monitor: observe, watch, keep an eye on
monoecious: having both the male and female reproductive organs in the same individual
monogamous: having only one mate at a time
monopolise: have complete and sole control over, take over control
monotreme/s: mammals who lay eggs and have no teats but secrete milk to feed their young
montane (geography) : specific ecological zones and ecosystems found at high elevations on mountains, particularly those below the timberline
morph (biology): one of the distinct forms or variations found within a species, especially when these variations are visually or behaviourally different
morphology, morphological: form of a thing
motivate. motivated: inspire, persuade, prompt, spark
moult: shed, cast, peel - when an animal sheds its old outer covering (skin/feathers, for example) to make way for a new one or new ones
mulga: an Australian shrub or small tree, *Acacia aneura*, forming dense growths in some areas
mysid shrimp: small, shrimp-like crustaceans
myth, mythology: belief, legend, lore, tradition
native: an animal or plant that has lived in a specific area for a long time, naturally, without being brought in from somewhere else
navigate, navigation: find a route to a destination
navigator: a person or animal who navigates
NCA: Nature Conservation Act 1992
nectar: sweet, sticky liquid made by flowers
nectarivore, nectarivorous: an animal which derives its energy and nutrient requirements from a diet consisting mainly or exclusively of the sugar-rich nectar produced by flowering plants
niche: a specialised area of interest, a position that is a perfect fit for someone, or a specific role an organism plays in its environment
nocturnal: active at night, relating to night
nomadic: wandering, travelling, roving
no-take: not allowed to be taken/caught
nucleus/nuclei: the structure in a cell that contains the chromosomes
nutrient/s: source of nourishment, food
nutrient-rich: rich in nutrients, nourishing food
nymph (insect): young insect that looks a bit like the adult, but smaller and without wings
oblanceolate (botany): having a broad rounded apex and a tapering base. simple, unsubdivided. (botany) of leaf shapes; of leaves having no divisions or subdivisions

obovate (botany): egg-shaped with the narrower end at the base. simple, unsubdivided. (botany) of leaf shapes; of leaves having no divisions or subdivisions
oceanic: marine, of the oceans
ocelli: simple eyes found on many insects and other invertebrates that detect light, but do not form images. They function as sensors for ambient light intensity, help with flight stabilisation, and are involved in orientation and movement detection
oesophagus: a muscular tube in the digestive system that carries food and liquids from the pharynx (throat) to the stomach
omnivore, omnivorous: eating both animal and plant foods
opportunistic: taking opportunities when they present
opposable, opposing: able to be placed opposite to something else
organelle: a **subcellular** structure that has one or more specific jobs to perform in the cell, much like an organ does in the body. Among the more important cell **organelles** are the **nuclei**, which store **genetic** information; **mitochondria**, which produce chemical energy; and **ribosomes**, which assemble proteins
organic: live, living, natural
organic material: matter that is made of organic compounds from the remains of organisms such as plants and animals, and their waste products
organism: animal, plant, structure, creature
osmeterium: a defensive, eversible (protrusible) organ found in swallowtail butterfly caterpillars, located just behind their heads, which they use to ward off predators by emitting a foul-smelling fluid and displaying bright, often red or orange, "horns"
ovary: the organ in a female animal that produces eggs and certain hormones; the part of a flower that contains the seeds and grows into a fruit
overexploit, overexploitation: overuse, excessive use of resources
overfish, overfished, overfishing: catching too many fish at once so that the breeding population is too reduced to be able to recover
overhunt, overhunted, overhunting: the act of hunting animals or birds to an excessive degree, which can be harmful to the population or the environment
overgraze, overgrazed, overgrazing: allowing grazing animals to reduce grazing land to beyond the point of recovery
ovum: an egg cell
ovule: an outgrowth of the ovary of a seed plant that after fertilization develops into a seed
paralysis: inability to move, can sometimes affect entire body systems such as breathing, movement, heart beat
parasite, parasitic: an organism that lives on or in an organism of another species (the host) from the body of which it obtains nutriment
parietal eye: a simple, light-sensitive organ on top of the head of some animals, like lizards and frogs, that acts as a "third eye"
passerine: relating to or denoting birds of a large order distinguished by having feet that are adapted for perching, including all songbirds
pastural/pastoralism: something related to rural life, especially the life of shepherds or country life
patagium/patagia *(pl.)*: wing membrane (bat), extensible fold of skin (insects, gliding mammals and reptiles), small process on the anterior thorax seen especially among butterflies and moths

pathogen: an organism or agent that causes disease, also known as an infectious agent or germ
pectoral fin: see fish anatomy diagram
peduncle (botany): the stalk of a flower
pelvic: having to do with the pelvis
pelvices: plural of pelvis
pelvic fin: see fish anatomy diagram
pelvis: the bony structure of the body below the abdomen that is located between the hip bones and contains the bladder and rectum
pentapedal: the use of the tail as a fifth leg during the slow gait of kangaroos
perch, perching: place where a bird can sit and rest, like a branch or a stick
perennial: lasting or existing for a long or apparently infinite time; (of a plant) living for several years
perianth (botany): the outermost, non-reproductive part of a flower, consisting of the calyx (sepals)and corolla (petals) when they are distinct, or tepals when they are undifferentiated
perimeter/s: boundary, edge, borderline
pesticide/s: a substance used for destroying insects or other organisms harmful to cultivated plants or to animals
petiole (botany): also known as a leafstalk, is the stalk that connects a leaf's blade to the stem of a plant
photosynthesis: how plants make their own food using sunlight, water, and air to produce oxygen and sugar, which they use for energy
phyllode (botany): modified petiole or leaf stem, which is leaf-like in appearance and function
phytoplankton: consisting of microscopic plants
piscivore/piscivorous: a carnivorous animal that primarily eats fish
pistil: the female reproductive part of a flower
placid: gentle, peaceful, quiet
plankton, planktonic: floating, drifting, or somewhat motile organisms occurring in a body of water, primarily comprising microscopic algae and protozoa
plastron: the ventral part of a turtle shell
plateau: a large, flat, elevated area of land that rises sharply above the surrounding terrain on at least one side
plume/plumage: feather (singular), feathers of a bird, collectively
poach/poaching: to illegally hunt or fish animals
poison, poisonous: likely to cause serious harm or death
pollen: a fine powdery substance, typically yellow, consisting of microscopic grain discharged from the male part of a flower or from a male cone. Each grain contains a male gamete that can fertilise the female ovule, to which pollen is transported by the wind, insects, or other animals
pollination: the act of transferring pollen grains from the male anther of a flower to the female stigma to help make new plants
pollinator, pollinators: animal, insect or other agent that transfers pollen to a plant allowing fertilisation
pollution: the introduction of harmful materials into the environment
polychaete worm: poly means 'many', chaete means 'hair', so these are a species of marine worm, commonly known as bristleworms.
polyclad, polyclads: any free-swimming, marine flatworm of the order Polycladida, having a broad, flat body and a many-branched gastrovascular cavity
polyoestrous: can raise several litters a year
population: number of inhabitants of an area
posterior (anatomy): behind, towards the back
precarious: dependent on circumstance, unsure, indeterminate, unreliable, doubtful
precipitation: what happens when water in the sky (clouds) falls back down to Earth as rain, snow, sleet, or hail

precocial: (of a young bird or other animal) hatched or born in an advanced state and able to feed itself and move independently almost immediately

predator/s, predation: an organism that exists by preying on other organisms

predict, prediction: to be able to tell in advance

pregnant/pregnancy: containing a developing embryo, foetus, or unborn offspring within the body

prehensile: adapted for seizing, grasping, or taking hold of something

premature: happening or done too soon, or before the expected or natural time

prey, preying: an animal, insect or organism hunted or taken for food, especially by a carnivorous animal

primary: most important, main

posture/posturing: behaviour that is intended to attract attention and interest

prokaryotes: organisms whose cells lack a **nucleus** and other **organelles**. **Prokaryotes** are divided into two distinct groups: the **bacteria** and the **archaea**, which scientists believe have unique evolutionary lineages. Most **prokaryotes** are small, single-celled organisms that have a relatively simple structure

probe/probing (birds): examining with the beak

prominent: standing out, obvious

protozoa, protozoan: any of a diverse group of eukaryotes that are primarily unicellular, existing singly or aggregating in colonies, are usually non-photosynthetic and often classified further into phyla according to their capacity for means of motility

protuberance/s: a rounded, outward projection or swelling on a surface

proximal (anatomy): closer to (root of the limb)

pseudo: false, fake, not genuine

puggle: baby echidna or platypus

pulmonate, pulmonates: having lungs or lunglike organs

pupa, pupae *(Pl.)***:** an insect in the nonfeeding, usually immobile, transformation stage between the larva and the adult stage

pupil (eye): the black opening in the middle of the coloured part of your eye

pursue: chase, track, follow, trail, hunt

pursuit: chase, hunt, quest, run

quarry (prey): something being hunted or pursued, specifically prey

quarry, quarrying: an open excavation usually for obtaining building stone, slate, or limestone

radio-tracking: a way of sending sounds or other information through the air from a transmitter attached to the animal to enable researchers to track it

rainforest canopy: the denser layer of greenery about 20 feet thick formed by the next tallest trees.

range (of species): an area where a member of a particular species can be found during its lifetime

rapid/rapidly: quickly, quick

raptor: usually a bird or prey, such as an eagle or a hawk

rare: not common or usual

recognise, recognition: seeing something you have seen before or can clearly identify by other means

recreation, recreational: fun activity, amusement, hobby

reflex: an automatic, involuntary, and nearly instantaneous physical response to a stimulus that does not require conscious thought

refuge: place that offers safety and protection from danger, trouble, or bad weather

regenerate/regeneration (biology): the process of renewal, restoration, and tissue growth that makes genomes, cells, organisms, and ecosystems resilient to natural fluctuations or events that cause disturbance or damage. Every species is capable of regeneration, from bacteria to humans.

remnant: remains, fragment, piece
reproduction/reproduce/reproductive: the production of offspring
reptile: cold-blooded vertebrate
residential development: housing development
respiratory system: breathing system, lungs
restrict, restricted: confine, restrain, regulate, limit
retina, retinae *(Pl.)*: inside back of eye of several layers
revere, revered: respect or admire greatly
reverse: backward, contrasting, opposite, contrary, change
reversible (physiological): something that can be changed from back to front or front to back
rhomboid, rhomboidal: parallelogram in which adjacent sides are not equal
ribosome: an intercellular structure made of both RNA and protein, and it is the site of protein synthesis in the cell
rigid: not able to be bent or changed easily, stiff
riparian: situated on the bank of water
ritual: a sequence of actions or words performed in the same order each time, often for a special purpose like a religious ceremony or a routine
riverine: of, relating to, or resembling a river. It can also describe something located or living on or near the banks of a river
robust: strong and healthy
rodent/s: a gnawing mammal of an order that includes rats, mice, squirrels, hamsters, porcupines, and their relatives, distinguished by strong constantly growing **incisors** and no **canine** teeth
rodenticide: a type of poison used to kill or control rodents
rods (eyes): specialised cells which are sensitive photoreceptors to light and help us see in low-light conditions. They are primarily responsible for night vision and peripheral vision, but they don't perceive colour
roost, roosting: perch where birds rest or sit
rufous: tinged with red, brownish red
rump: hindquarters, buttocks, sacral region of a bird
rural: relating to the countryside and not to towns or cities
sagittal (anatomy): dividing body into right and left
saline: containing or impregnated with salt
salinity: saltiness
samara: a dry indehiscent usually one-seeded winged fruit
sanctuary/sanctuaries: a place of refuge or protection, offering safety from danger or hardship
satellite: an artificial body placed in orbit round the earth or moon or another planet in order to collect information or for communication.
satellite tracking: using satellites orbiting earth to track
satellite transmitter: device which sends a signal to a satellite
saturate, saturated: soaked, soggy, waterlogged
savannah: flat grassland, plain
scarce, scarcity: not plentiful, hard to find or hard to get
scavenge/scavenger: an animal that feeds on the remains of other animals or eats decomposing plant or animal matter. This means that scavenger animals can be both herbivores and omnivores. Most times, scavengers do not kill their own prey, even if they may possess the ability to do so
school, schooling (of fish): group of fish moving together
sclerophyll forests : vegetation that has adapted to long periods of dry, hot weather
scrubland: a plant community characterised by vegetation dominated by shrubs
sea squirt: invertebrate, usually fixed potato-shaped organism using sac-like filter feeding
secretive: keeping secrets
sedentary: staying in one place, not moving much

sedge: plant that looks like grass, but it grows in wet areas
sediment, sedimentation: organic matter settling to bottom of a liquid
sedimentary: of or relating to rocks formed by the deposition of sediment
segment/s: part of, divided parts
self-pollination: the pollination of a flower by pollen from the same flower or from another flower on the same plant
semi- (prefix): half, partially, incompletely
semi-arid: climate that has little rain but is not completely dry
semi-parasitic: plant which gets water or food from host but can also produce its own food
semi-prehensile: able to grasp but cannot really hang on tightly
semi-transparent: almost see through
sepal: the green, leaf-like part of a flower that protects the bud before it blooms and supports the petals when the flower is open
serotiny: following, later - when plants retain seeds in cones or fruits until fire, which melts resins, releasing them
serrated: having a jagged or toothed edge, like a saw
seta/setae (*Pl*): a stiff, bristle-like hair or process on an organism, with different meanings depending on the context. In biology, it can refer to sensory and locomotive hairs on invertebrates like spiders and earthworms, the stalk of a moss capsule, or a bristle on a caterpillar
settlement: a place, typically one which has previously been uninhabited, where people establish a community
shellfish: seafood like molluscs, crustaceans
sheoak: type of tree
sibling: each of two or more children or offspring having one or both parents in common; a brother or sister
sift: an act of sifting something, especially so as to isolate that which is most important
significant, significance: very important or has a special meaning
silt: fine sand, clay or other material carried by running water
siltation: water pollution by silt
skim, skimming: to pass lightly or hastily; glide or skip along, above, or near a surface
snout: the long nose and mouth of an animal that sticks out
snout-to-vent: a measurement of the body length of an animal, specifically from the tip of the snout to the cloacal opening, or vent
social: friendly, like company
solitary: by itself, single
sophisticated: complex and intricate
spawn, spawning: reproduce, generate (usually fish or frog eggs)
sparse: something that is thinly spread out or not dense
species: one of the classes of things included with other classes in a genus
specific:clear-cut, particular, precise, unique
specimen: sample, example
sperm: the male reproductive cell, also known as a gamete, which contains the father's genetic material and is responsible for fertilising a female egg to begin the development of a new individual
spinifex: grass plant with stiff, sharp-pointed leaves that grow in dense masses
spinose: having many sharp, pointed things sticking out; spiny, prickly
spiracle/s: the opening in the exoskeletons of insects, myriapods, velvet worms and many arachnids to allow air to enter the trachea
spire: a part of the coiled shell of molluscs. The spire consists of all of the whorls except for the body whorl
spoil heap: an untidy pile or mass usually of refuse material from an excavation

spore: a reproductive cell that can develop into a new individual without uniting with another reproductive cell
spur (biology): an outgrowth of bone covered in a sheath of horn found in various anatomical locations in some animals
staccato: detached, short, and abruptly disconnected
stalk (hunt): walking quietly in search of animals or in pursuit of an individual animal
stalling: stopping (flying)
stance: way of standing physically or figuratively
stamen: the male part of a flower that makes pollen
stigma/s: the part of a flower that receives pollen and is the tip of the **pistil**
strident: loud and harsh, or describes a forceful, insistent, and often irritating way of expression
stolon (botany): a creeping horizontal plant stem or runner that takes root at points along its length to form new plants
stochastic: random or involving chance
storm front: border between two different kinds of air, often with different temperatures; when a front is created, weather such as storms and **precipitation** happen
streamers (tail feathers): tail streamers on birds are the elongated narrow tips of the tail feathers; elongated, specialised tail feathers that are often used for display and communication, particularly during mating rituals. They can be quite long, sometimes exceeding the bird's body length, and are often vibrant in colour or pattern
streamline, streamlined: designed to have the least resistance
style (botany): of an angiosperm flower is an organ of variable length that connects the ovary to the stigma
sub- (prefix): underneath, below, less than
subcellular: of less than cellular scope or level of organisation
subfossil: the remains of an organism that are partially fossilised, meaning they are neither fully modern nor completely mineralised like a true fossil
submissive, submission: the act of accepting control by another
subordinate: a person or thing that has a lower rank or is less important than another
subpopulation: specific small section within a larger group
subspecies: additional species within a grouping
substrate: the material or substance an organism lives on or attaches to, or it's the molecule that an enzyme acts upon in a chemical reaction
subtropical: bordering on the tropics; nearly tropical. pertaining to or occurring in a region between tropical and temperate; semitropical
superficial (anatomy): on the surface
superior (anatomy): above, higher than
supra- (prefix): above, beyond, or earlier in time
supralittoral: area above shoreline, likely damp
survive, survival: stay alive, staying alive
susceptibility: easily affected, harmed, or influenced by something
swamp: an area of low-lying, uncultivated ground where water collects; a bog or marsh
swamp drainage: emptying water of swamp
swim bladder: special organ inside a fish that helps it float and control its depth in the water
sygnathids: a large and diverse group of fishes, including seahorses, pipefishes, seadragons and pipehorses
symbiotic, symbiosis: living together in a mutually beneficial way

tail streamers: tail streamers on birds are the elongated narrow tips of the tail feathers; elongated, specialised tail feathers that are often used for display and communication, particularly during mating rituals. They can be quite long, sometimes exceeding the bird's body length, and are often vibrant in colour or pattern

talon: the sharp, curved claw on the foot of a bird of prey, used for hunting and gripping prey

tannin: astringent vegetative compound, reddish/brown

tassel, tasselled: fringed lobes of skin similar to tassels, which are made of multiple cords, usually hanging down

tawny: dark or dull yellowish-brown

temperate: moderate in respect to temperature

tendril: thin organs of (usually of plants) that wrap around other objects for support

termitarium/termitaria: termite nest/s

terrestrial: land based, growing or living on land

territory, territories, territorial: a tract of land, somewhere that the animal recognises as their territory which they will protect if they are territorial

testes: an organ which produces spermatozoa (male reproductive cells)

thermal: using, resulting from, or producing heat

thermal current: energy that is transferred from one body to another because of a difference in temperature creating a current

thicket: a dense growth of bushes or small trees that are packed closely together, often making it difficult to pass through

thorax: the portion of the body between the head and the abdomen

threat: alarm, danger, hazard, menace, peril, risk, warning

threatened: in danger or likely to be harmed, or to become more significantly vulnerable or endangered

timid: shy, nervous, and lacking confidence, or it can mean overly cautious and hesitant due to fear

tissue: a group of similar cells and their intercellular material that work together to perform a function

tolerate: put up with

torpor: a state of physical or mental inactivity; lethargy.

totem, totem animal: a natural object or animal that is believed by a particular society to have spiritual significance and that is adopted by it as an emblem

toxic/toxin/toxucity: poison, poisonous

trait: a specific characteristic or quality of an animal, or thing that makes it unique

trample. trampled: when hard-hooved animals stamp or tread heavily on the ground

translocation: the movement of something from one place to another

transmit, transmitted: to send or pass something from one place or person to another, like a message or a disease

transparent: see-through, light passes through completely like clear glass

transverse (anatomy): dividing body into upper and lower sections

trauma, traumatic: frightening, dangerous, or violent event that poses a threat to life or bodily integrity; wound or injury which can be both physical and/or psychological, or both

trawl: (myotis) fish by dragging feet through water

tree canopy: the upper parts of a tree, leaning into neighbouring trees

trenches (ocean zone): hadal zone

tropics, tropical: regions of Earth closest to the imaginary line called the equator, which runs around the middle of the planet. It's always warm in the tropics, with high humidity and plenty of rain

truffle: a fungus that grows underground and is edible

truncate: to cut off short; abbreviate, cut, the movement of something from one place to another, foreshorten, shorten

trypanosome: genus of **kinetoplastids**: a monophyletic group of unicellular parasitic flagellate protozoa

tuber/s: underground part of a stem or rhizome, like potatoes or sweet potatoes.

tubercles, tuberculate: a little bump or knob, kind of like a tiny wart, that can grow on plants or on the body, especially on bones

tundra: major zone of treeless level or above the timberline on high mountains, rolling ground found in cold regions, mostly north of the Arctic Circle

turbulent/turbulence: characterised by conflict, disorder, or rapid, uneven changes. It can describe situations, periods of time, or even physical movements like air or water flows. Think of it as the opposite of smooth, stable, or calm.

tympanum: the eardrum or tympanic membrane

ultraviolet light: a type of electromagnetic radiation with a wavelength shorter than visible light but longer than X-rays, making it invisible to the human eye

understorey: the layer of plants and especially the trees and shrubs between the forest canopy and the ground cover

unicellular: consisting of a single cell

unique, uniqueness: one of a kind, unlike anything else

urban: relating to a particular place, town, village, in this instance development due to humans needing to have more homes

uterus: a hollow, muscular, pear-shaped organ in the female pelvis where a baby develops during pregnancy

vacuum: a space where most of the air or matter has been removed, creating a void

vagrant: wandering from place to place

van der Waals' forces: weak, short-range attractive or repulsive forces between molecules or atoms that are not the result of a chemical bond. Geckos use van der Waals forces to climb walls by creating many small attractions between the hairs on their feet and the surface

vary, variable, variation: the quality or state of being different or diverse, not uniform

vegetarian: eating only plant matter

vegetation: plants

vehicular traffic: movement of cars, trucks, and other vehicles on roads

venom, venomous: poison, poisonous

venom glands: the specialised glands responsible for producing the poisonous secretions of an animal, often associated with spines, teeth, stings, or other piercing devices

vent (anatomy): the external opening of the **cloaca**, a single posterior orifice that serves as the only opening for the digestive, urinary, and reproductive tracts in many animals, including birds, reptiles, amphibians, and some fish and mammals

ventral (anatomy): towards the front

ventrolateral/ventrolaterally: the front and side of a body or body part

verify, verified, verification: prove the truth of, evidence

vertebra, vertebrae, vertebrate: spinal column or backbone

vertical: upright, top to bottom

vibrate/vibratory: to move back and forth or up and down rapidly and rhythmically, like shaking or quivering

vicinity: the area around something

victor: a conqueror or winner

vigorous/vigorously: showing a lot of energy, strength

virus: an infectious microbe consisting of a segment of nucleic acid (either DNA or RNA) surrounded by a protein coat. A virus cannot replicate alone; instead, it must infect cells and use components of the host cell to make copies of itself

vivid: strikingly bright or intense
vocalisation/s: the act of producing sounds with the voice
void: empty space
voracious: having a huge appetite; ravenous
vulnerable: weak, unsafe, defenceless, sensitive
warren: a network of interconnected burrows
wary/wariness: cautious and alert to potential danger
water hole: a natural puddle or small lake in a dry area where animals go to drink
water-logged: saturate with water
water table: the boundary between saturated and unsaturated ground
weed: plant growing where it's not wanted
weir/s: bank, barrier, ditch, dike
wet sclerophyll forest: characterised by very tall eucalypt trees usually in moist cloudy uplands with up to 2500mm of rainfall each year
wetlands: an area of land that's covered by water or has waterlogged soil, meaning the ground is full of water, like a sponge
whorl/s: something that swirls or winds around like a spiral
wingspan: the maximum extent across the wings of a bird or other flying animal, measured from tip to tip
xanthophore/s: a **chromatophore** containing a yellow pigment that is typically a **carotinoid** and occurring especially in fishes and crustaceans

NOTES

REFERENCES AND CITATIONS:

1. IUCN (2024, July 19). *IUCN Red List*. International Union for the Conservation of Nature. Retrieved August 8, 2024, from https://www.iucnredlist.org/
2. Noongar - Wikipedia. https://en.m.wikipedia.org/wiki/Noongar
3. Black-tailed godwit (Limosa limosa) - Wild travel in Danube Delta. https://wildtravel.ro/bird/black-tailed-godwit-limosa-limosa/
4. Banded hare-wallaby - Wikipedia. https://en.wikipedia.org/wiki/Banded_hare-wallaby
5. Rainbow Bee-eater | BIRDS in BACKYARDS. https://www.birdsinbackyards.net/species/Merops-ornatus
6.

To claim your Conservation Club reward badge for NOONGAR COUNTRY:

- **Answer the following 20 questions.**
- **Send either a photograph or a scan of one colouring page** which you have coloured in reasonably correctly as far as colours of your chosen species is concerned!

* **Postage will apply both in Australia and overseas. This will be estimated and sent to you for approval via email prior to shipping.**

Let's see what you have learned, whether you are 8, 80, or somewhere in between! There are clues which are usually underlined or easily found throughout the book.

1. How many species can you find in the book that are EXTINCT according to the IUCN (International Union for Conservation of Nature)?
2. Why is the sunset frog called the sunset frog?
3. How many hearts does an octopus have?
4. Which species of lizard in the book is known to have a venomous bite?
5. What names are a group of pelicans called?
6. Which bird in the book has a long, upturned black bill?
7. Which tiny insect is named after a bird with a spectacular tail feather display because it also has a very colourful mating display?
8. Which lizard species has a fake head on its neck used for deterring predators?
9. Which bird is very important for spreading the seeds of mistletoe plants?
10. Which bird gets its name because of its habit of flipping over stones when searching for food?
11. Which species of marsupial is also known as a Boodie, the Noongar name for this animal?
12. Which species of marsupial had a claw-like spur on the end of its tail?
13. Which species of bird is the official state bird of Western Australia?
14. Which tiny species of marsupial feeds entirely on nectar and pollen?
15. Which species of bird is a pretty terrible nest builder but very good at camouflaging itself in tree branches?
16. What insect species featured in the book was a critical food source for Indigenous peoples?
17. Which marine species has "walking" fins and can make loud grunting sounds?
18. Which species of bird has been recorded to fly more than 48 hours without stopping during migration?
19. Which species of tiny plant featured in the book is a carnivorous plant, capturing and digesting insects?
20. Which bird species lays eggs approximately the size of a pea?

INDEX

Made in the USA
Monee, IL
08 November 2026

68085577R00096